Caution: Faulty Thinking Can be Harmful to Your Happiness

Logic for Everyday Living

Third Edition

Elliot D. Cohen Ph.D.

CENGAGE Learning™

Australia • Brazil • Japan • Korea • Mexico • Singapore • Spain • United Kingdom • United States

CENGAGE
Learning™

Caution: Faulty Thinking Can be Harmful to Your Happiness
Logic for Everyday Living
Third Edition

Elliot D. Cohen Ph.D.

Executive Editors:
Michele Baird
Maureen Staudt
Michael Stranz

Project Development Manager:
Linda deStefano

Senior Marketing Coordinators:
Sara Mercurio
Lindsay Shapiro

Production/Manufacturing Manager:
Donna M. Brown

PreMedia Services Supervisor:
Rebecca A. Walker

Rights & Permissions Specialist:
Kalina Hintz

Cover Image:
Getty Images*

* Unless otherwise noted, all cover images used by Custom Solutions, a part of Cengage Learning, have been supplied courtesy of Getty Images with the exception of the Earthview cover image, which has been supplied by the National Aeronautics and Space Administration (NASA).

ISBN-13: 978-0-759-33630-8

ISBN-10: 0-759-33630-X

Cengage Learning
5191 Natorp Boulevard
Mason, Ohio 45040
USA

Cengage Learning is a leading provider of customized learning solutions with office locations around the globe, including Singapore, the United Kingdom, Australia, Mexico, Brazil, and Japan. Locate your local office at:
international.cengage.com/region

Cengage Learning products are represented in Canada by Nelson Education, Ltd.

For your lifelong learning solutions, visit **custom.cengage.com**

Visit our corporate website at **cengage.com**

Printed in the United States of America

CONTENTS

INTRODUCTION 1

PART 1:

HOW TO IDENTIFY AND AVOID FAULTY THINKING:
A DIRECTORY OF FALLACIES, 7

CHAPTER ONE:
RATIONAL THINKING IS A GUIDE TO HAPPINESS, 9

Three Reasons Why the Way You Think Matters, 9
1. *Your thinking shapes and molds your view of reality,9*
2. *Your thinking largely determines your feelings and emotions, 10*
3. *Your thinking influences your behavior, 12*

Thinking As A Primary Human Responsibility, 12

Fallacies: What Are They and Why Avoid Them?, 13

Some Tips On Finding Fallacies In Your Thinking, 14
1. *Getting at fallacies by asking "Then What?", 15*
2. *Getting at fallacies by philosophically asking "Why?", 15*
3. *Getting at fallacies by asking "Why is that a reason?", 16*

About the Practice Scenarios, 16

Software Applications, 17

1. *BELIEF-SCAN: Artificial Intelligence for Detecting and Diagnosing Faulty Thinking, 17*
2. *Suggested Practical Exercises Using BELIEF-SCAN, 19*

Thinking About Your Own Thinking:
Keeping a Personal Log, 21

Improving Your Thinking Skills:
Some Preliminary Considerations, 23

Benefits of Avoiding Faulty Thinking:
Some Testimony, 26

CHAPTER TWO:
FALLACIES OF FACT GATHERING, 29

What are the Fallacies of Fact Gathering?, 29

1. *The Straw Man, 29*
2. *Slippery Slope, 30*
3. *Magnifying Risks,30*
4. *Half Truth, 31*
5. *The "Had Things Been Different" Fallacy, 32*
6. *False Comparison,33*
7. *Unsupported Explanation, 33*
8. *Oversimplifying, 35*
9. *Hasty Generalizing, 35*
10. *Stereotyping, 36*
11. *Ignoring the Past, 38*
12. *Insisting on the Past, 38*
13. *Assuming the Cause, 39*
14. *Questionable Claim, 39*
15. *The Smoke Screen, 40*

Review Definitions:
Fallacies of Fact Gathering, 41

Practice Scenarios
For Chapter Two, 43

Further Practical Exercises
For Chapter Two, 55

CHAPTER THREE:
LOGICAL FALLACIES, 57

What Are Logical Fallacies?, 57
1. *The Fallacy Of The Whole, 57*
2. *The Fallacy Of The Part, 58*
3. *Arguing In A Circle, 58*
4. *The False "Either Or", 59*
5. *Gray Area Fallacy, 60*
6. *Inconsistency, 60*
7. *The Vice Versa Fallacy, 62*
8. *Fallacy Of Deriving "If" From "Then", 62*
9. *Fallacy Of Deriving "Then Not" From "If Not",64*

Review Definitions:
Logical Fallacies, 66

Practice Scenarios
For Chapter Three, 67

Further Practice Exercises
For Chapter Three, 75

CHAPTER FOUR:
FALLACIES OF MEANING, 77

What Are Fallacies of Meaning?, 77
1. *Personifying, 77*

2. *Switching Meanings, 78*
3. *Vague Terms, 78*
4. *Opposite Meaning, 79*
5. *Self-Evidence Fallacy,80*

Review Definitions:
Fallacies of Meaning, 81

Practice Scenarios
For Chapter Four, 82

Further Practical Exercises
For Chapter Four, 87

CHAPTER FIVE:
RELEVANCE FALLACIES, 89

What Are Relevance Fallacies?, 89
1. *Appeal to Ignorance, 89*
2. *Appeal to Force, 90*
3. *Two Wrongs Make a Right, 90*
4. *Misuse of Authority, 91*
5. *Jumping on the Bandwagon, 91*
6. *Parroting, 91*
7. *Appeal to Tradition, 93*
8. *Misuse of Pity, 93*
9. *The World Revolves Around Me, 94*
10. *Sweeping Generalization, 95*
11. *Personal Attack, 95*
12. *Changing the Subject, 96*
13. *Hairsplitting, 97*

Review Definitions:
Relevance Fallacies, 99

Practice Scenarios
For Chapter Five, 100

Further Practical Exercises
For Chapter Five, 115

CHAPTER SIX:
FALLACIES OF MAKING VALUE JUDGMENTS, 117

What Are Fallacies of Making Value Judgments?, 117
1. *Demanding Perfection (In This Imperfect World), 117*
2. *I-Can't-Stand-It-It is, 118*
3. *Awfulizing, 119*
4. *Slippery Slope-awfulizing-I-Can't-Stand-It-It is Syndrome,120*
5. *Damnation, 120*
6. *Poisoning The Well, 121*
7. *The "Terrific" Fallacy, 122*
8. *The " I Just Can't Help This Feeling" Fallacy, 122*
9. *The "Thou Shalt Upset Yourself" Fallacy, 123*

Review Definitions:
Fallacies of Making Value Judgments, 124

Practice Scenarios
For Chapter Six, 125

Further Practical Exercises
For Chapter Six, 134

<u>PART 2:</u>

KEEPING YOUR WITS ABOUT YOU: *RATIONAL THINKING AND EMOTION,137*

CHAPTER SEVEN:
ASSERTING YOURSELF, 139

Avoiding Fallacies As A Goal Of Assertiveness Training, 139

Seven Ways To Overcome Inertia, 139

1. *Put your actions where your mouth is, 139*
2. *Be especially on the lookout for Damning thoughts about yourself, 140*
3. *Don't lock yourself into inertia for fear of how you think others might view you if your efforts were to fail, 141*
4. *Please yourself, 142*
5. *Learn how to say no, 143*
6. *Avoid making your decisions by indecision, 144*
7. *Avoid dilemma thinking, 145*

Practice Scenarios
For Chapter Seven, 146

CHAPTER EIGHT:
REDUCING YOUR STRESS, 153

Stress As Intense Emotion, 153

Anxiety, Anger, Depression and Guilt: 155

What Fallacies To Look For In Each,

1. *If you are experiencing intense anxiety.... 155*
2. *If you are experiencing intense anger....156*
3. *If you are feeling depressed....157*
4. *If you are feeling intense or debilitating guilt....159*

Six Additional Ways to Reduce Stress, 161

1. *Do what you can about the problem, 162*
2. *Distract yourself, 162*
3. *Schedule yourself, 163*
4. *Don't take on more than you can handle,164*
5. *Avoid rating yourself, 164*
6. *Accept yourself as an imperfect person, 165*

Practice Scenarios
For Chapter Eight, 166

APPENDIX A:
Your Faulty Thinking Prognosis:
Fallacies You May Still Need to Work On, 174

APPENDIX B:
Suggested Answer to Starred Practice Scenarios, 176

Further Reading, 187

INTRODUCTION

Does there appear to be a lack of communication between you and close relations (parent, partner, child, etc.)? Do you have trouble making effective decisions? Are you frequently stressed out or anxious? Do you get extremely anxious at certain times such as before an exam or job interview? Do you feel like certain circumstances or persons are `driving you crazy'? Do you have difficulty controlling your temper, get frustrated or upset easily? Do you have difficulties getting along with others? Are you and your mate frequently embroiled with one another? Are you depressed over some event in your life? Do you find it hard to assert yourself? Do you sometimes feel like you are losing control over your own life?

While the causes and explanations of such life difficulties are quite complex, their roots can often be traced to faulty thinking, to bad logic. Whether you realize it or not, you make hundreds of inferences every day. When enough of this logic is bad, your life can suffer in precisely the sort of ways mentioned above.

Moreover, bad logic has a cumulative effect. Faulty thinking today leads to decisions, actions, beliefs, emotions, and attitudes that can send you into a tailspin tomorrow, which can then affect you the next day, and so on. After many years of thinking illogically you can well imagine the damage that is done to your happiness. So, as the title of this book clearly states, *Caution: Faulty Thinking Can Be Harmful to Your Happiness*!

In our medically advanced society, many people are concerned about their physical health. They avoid high cholesterol and are conscious of the importance of exercise and, in general, good physical hygiene. They rightly realize that their physical well-being is needed for their overall personal happiness. Unfortunately, you don't often find people worrying as much about problems in their process of thinking. While they watch out for cholesterol, they are much less cautious about watching out for bad logic. Yet a steady diet of faulty thinking can also have very serious negative effects on your life.

The logic in this book is an *everyday* one, which means it can be used to deal with the events in your live that occur from day to day. It is a down to earth practical one and it avoids unnecessary theory and technical jargon. This makes its content accessible to many. However, while it is non-technical and practical, it is based upon very solid, well tested, principles and theories.

It is inspired and infused by two primary sources. The first is a growing trend in higher education, as well as in secondary and elementary education, toward improving reasoning skills in students. This movement has generated much research and development of texts in applied logic for classroom use.

The second source is a trend in current psychology toward what are usually called cognitive-behavioral theories, especially the theory of Albert Ellis called "Rational Emotive Therapy." Having bases in Stoic philosophy, these theories are concerned with the psychological impact of irrational thinking on your ability to cope satisfactorily with life situations. These theories attempt to classify kinds of irrational thinking and to use behavioral techniques as ways of helping people to solve their problems of living. Two works of this nature that have had a noteworthy influence on the direction of this book are Albert Ellis, "The Philosophical Basis of Rational-Emotive Therapy (RET)," *International Journal of Applied Philosophy*, Vol. 5, No.2 (Fall 1990); and Albert Ellis and Robert A. Harper, *A New Guide To Rational Living* (No. Hollywood, Cal: Wilshire Book Company, 1975). Many of the fallacies introduced in Chapter 6 are credited to Ellis.

However, while most applied logic texts today tend to ignore these insights of psychology, the psychological literature similarly tends to ignore the insights of applied logic. Although the psychological literature classifies types of irrational thinking, it typically does not cover the myriad of practically relevant distinctions made in applied logic texts (which are usually written by philosophers who have distinct professional training in the area of logic).

On the other hand, while the applied logic texts aim at addressing practical matters, they usually do not adequately address those types of *private* issues that occupy the greater part of people's lives such as interpersonal and personal issues in marriage, family, friendship, dating, work, recreation, and school. Thus, for example, many examples of how logic "applies" are borrowed from the *public* sector such as advertising, politics, and mass media sources. (While such applications are not the primary focus of the present treatment, practical exercises concerning these have also been included.) Moreover, the focus of such texts is typically on finding fallacies committed or invited by the authors of these publications, not on examining the fallacies inherent in your *own* thinking.

In the past two decade or so, there has been a growing current in applied philosophy known as philosophical counseling or practice, which focuses on putting philosophy to work to help people solve problems of everyday life. While some philosophers have claimed this field to be distinct from psychological counseling, I have avoided this claim. The distinguishing feature of philosophical counseling is rather its greater emphasis on philosophical methods and theories in helping clients to solve their behavioral and emotional problems. To the extent that the training of philosophers and psychologists is distinct, it is obvious that their approaches will vary.

As I have argued in my recent book, *What Would Aristotle Do? Self-Control Through the Power of Reason* (Prometheus, 2003), it is philosophy's methods of logical and critical thinking that largely account for its practical power. To the extent that philosophical counseling harnesses philosophy to help ordinary people think more logically, critically, and, hence, more realistically about their problems of living, it has inestimable value. My own clinical research over the past two decades has borne this out.

Caution: Faulty Thinking Can Be Harmful To Your Happiness brings some of the most important insights of both philosophical counseling and psychology into its development of applied logic. Focusing primarily on problems and issues of everyday living from the perspective of self-examination and self-improvement, it is able to penetrate to a deeper practical level than any usual treatment.

To further serve the purposes of self-examination and self-improvement this book is available with supporting software designed explicitly for these purposes: *BELIEF-SCAN FOR WINDOWS: ARTIFICIAL INTELLIGENCE FOR DETECTING AND DIAGNOSING FAULTY THINKING.*

This text may be used in courses on critical thinking, informal logic, decision making and a variety of other courses that emphasize the role of rational thinking in satisfactory coping, including courses on assertiveness, stress management, and employee training programs. It can be used effectively in individual as well as group contexts to help people deal with the myriad of problems (including emotional ones) that are rooted in faulty thinking. This book has been used to provide instruction for diverse populations including, for example, college students, high school students, recovering drug and alcohol addicts, displaced homemakers, secondary school teachers, nurses, students at risk, and inmates in correctional institutes.

Caution: Faulty Thinking Can Be Harmful To Your Happiness is organized into two main parts. The first part, called "How To Identify and Avoid Faulty Thinking: A Directory of Fallacies," names, discusses and illustrates different varieties of faulty thinking or "fallacies." It also contains a number of practice exercises and assignments that will help you to avoid faulty thinking in daily living. Among other things, Chapter 1 discusses these exercises in greater detail (those referred to as "Practice Scenarios') and it tells you how to attain maximum personal benefits out of the use of this book. You should read Chapter 1 before going on to subsequent chapters.

The second part of this book, "Keeping Your Wits About You: Rational Thinking And Emotion" gives you specific directions for using what you have learned in the first part of the book to deal effectively with problems of inertia (lack of assertiveness and the failure to take charge of your own life). It also targets four common forms of emotional stress (anxiety, anger, depression and guilt).

This book is a product of almost two decades of teaching, research and field work in everyday logic, which has been conducted in university, community college, adult education, workshops, and clinical settings. While many have benefited from its ideas, there is no *guarantee* that it will help you. As is the case with any instructional tool, your successful use of it will depend upon a number of variables. For example, some "thinking" problems may require the care of a physician or psychiatrist, while others may require the care of a psychologist, mental health counselor, marriage and family counselor, social worker or other appropriate mental health specialist. The ideas and suggestions contained in this book are not intended as a substitute for the consultation and/or care provided by these specialists.

The logic developed in *Caution: Faulty Thinking Can Be Harmful To Your Happiness* is dynamic and systematic. It will be most effective if you work through all the chapters, beginning with the first chapter. It will be least effective if you work through it in a piecemeal fashion. For example, do not attempt Part 2 without having carefully worked through Part 1 first. However, if you have problems rooted in faulty thinking (as all of us do to one extent or another), then your efforts are likely to pay off.

PART 1

HOW TO IDENTIFY AND AVOID FAULTY THINKING:

A DIRECTORY OF FALLACIES

CHAPTER ONE
RATIONAL THINKING AS A GUIDE TO HAPPINESS

THREE REASONS WHY THE WAY YOU THINK MATTERS

There are three important effects that your thinking can have on you, which can explain how thinking is able to affect your happiness. Let's briefly consider them:

1. Your Thinking Shapes And Molds Your View Of Reality

An important idea to keep in mind is that your perception of reality does not occur independently of your own thinking. Rather, your thoughts shape and mold your perception of reality. Moreover, there are at least two ways in which your thinking can affect your perception of reality:

First, much of what people call reality is arrived at by a process of thinking called *inference*. An inference is made when you attempt to come to a conclusion about something on the basis of something else that you already know or think you already know. When you make inferences you are also doing what is called *reasoning*. For example, if you notice that your mate is unusually quiet, you may infer, or reason, from this information, that he/she is angry or displeased with something you've done. If you made this inference on just that information, your inference would be an irrational one (as you will see in Chapter 2). Indeed, another alternative inference you could just as easily make from the very same information would be that he/she is preoccupied with some matter in his/her life that is not related to anything you have done. In either case, however, notice how your interpretation of reality would be affected by the inferences you make. More generally, keep in mind that what you take to be reality consists, to a large extent, of things that you have inferred, and not merely of what you have actually observed.

Second, the *value judgments* you make about events in your life shape and mold your view of reality. For example, suppose you lose your job. You could interpret this event as "the worst thing that could ever have happened." As you will see (in Chapter 6) this would normally be a very irrational interpretation. Nevertheless if you *did* interpret this event in this way, then you would undoubtedly focus on what you took to be the negative aspects of your situation, for instance, "I'll never get another job like this again" and "I'll probably end up in the gutter." But you could, alternatively (and more rationally), interpret this event as unfortunate or even as a good thing, depending upon whether or not you really *wanted* the job in the first place! In this last case you would undoubtedly perceive different and more positive aspects of your situation, for instance, "Now I'll be able to find a job that I like better." Notice how the different value judgments you make about what has happened to you influence and direct your interpretation of reality—as well as how you *feel* about it.

2. *Your Thinking Largely Determines Your Feelings And Emotions*

In making value judgments about the events in your life you can also determine your own *feelings or emotions* about those events. For example, in telling yourself that the loss of your job is "the worst that could happen," you would normally upset yourself. Notice how hard it would be for you to tell yourself this without, at the same time, feeling disturbed! More generally, keep in mind that your feelings and emotions are normally involved in your value judgments. For instance, if you are angry at somebody, then you probably have decided that he/she has done something wrong to you. If you are *furious* at him/her, then you probably think that what he/she has done to you was terrible, horrible and awful. For this reason, avoiding intense and self-defeating feelings and emotions may mean re-evaluating the events in your life in more rational ways. For example, instead of thinking that the loss of your job was the worst that could happen, you could more rationally decide that it was unfortunate, but nevertheless not the worst thing in the world. In this way, you would probably not feel quite so bad.

In addition to value judgments, your emotions generally include other types of related thinking. In particular, when you make a value judgment, for example, "It's awful that my husband divorced me," you will generally base your judgment on certain *factual* claims, for example, "Now I'm going to end up alone for the rest of my life, all by myself." It is thus such a network of *value-judgments-supported-by- factual-claims* that largely determines your emotions, for example in this last case, depression.

Moreover, by carefully reviewing and examining your factual support for your value judgment, it is in *your* power to change your value judgment. For example, by coming to see that you need not end up "all by yourself" as a result of your divorce, and that there are many other new options open to you, it is in your power to re-evaluate your situation, and, thereby, to change your emotion itself. For example, by rethinking the situation, you can move from depression to a much milder emotion, say regret: "Well, it's too bad it had to end this way after all we've been through together, but there is still a world out there with many options for me to explore." Compare the *feeling* of telling yourself this to that of telling yourself how *horrible* it is to be divorced! Indeed, by speaking to yourself differently you can, in fact, decide to feel one way or the other.

3. Your Thinking Influences Your Behavior

Your *value-judgments-supported-by-factual-claims* are also linked to your behavior. Indeed, your behavior is often a concrete expression of that thinking. For example, if John thinks that his wife is deliberately giving him "the cold shoulder," say merely because she is exceptionally quiet, he may well *act* defensively toward her, especially if he thinks that "Nobody must ever treat me like that," and "She's no damn good for treating me like that." For example, he may "tell her where to go" or he may deliberately try to avoid her. However, because John's actions are the result of irrational reality inferences and value judgments, they are likely to be self-defeating, leading exactly to what John would not welcome, namely, his wife's dissatisfaction with him.

Since it is in your power to re-interpret situations in your life, it is also possible to *re*-direct your own behavior. That is, you can change your behavior by changing your thinking. For example, it is in John's power to consider other explanations for why his wife is quiet. After all, maybe there is something on her mind quite unrelated to John. And it is also in John's power to re-*evaluate* his situation *even if* it is true that his wife is angry at him for some reason or other. "Well, I would prefer it if she would be more up front with me when she's angry at me; but the fact that she isn't does not mean that she's do damn good." In so changing his thinking, John will be less inclined to engage in self-defeating behavior toward his wife and, instead, more inclined to explore constructive avenues such as coolly talking things over with her.

THINKING AS A PRIMARY HUMAN RESPONSIBILITY

It should now be clear precisely why the way you think is extremely important for your happiness. It is because your thinking shapes and molds your view of reality, largely determines your emotions, and influences your behavior. If your view of reality is misguided as a result of twisted thinking, you are not likely to accomplish your goals or to have attainable expectations;

if your emotions are the products of faulty thinking, you are not likely to form a healthy self concept or to relate well to others; if you act without thinking clearly and rationally about what you are doing, you are likely to make decisions that frustrate your own efforts and lead to regrets.

However, your thinking is not merely something that happens *to you*, like a bolt of lightening out of the blue. What you think is itself largely in *your* power; and, given the far reaching effects of your thinking on your own happiness--and undoubtedly upon that of others--*taking responsibility for your own thinking should be viewed as a primary human responsibility*, perhaps the most important one you have. So what can be done to tune up your thinking?

FALLACIES:
WHAT ARE THEY AND WHY AVOID THEM?

In this book you will learn to detect and avoid *fallacies* of thinking. *A fallacy can be defined, for our practical purposes, as* ***a way of thinking or reasoning that has a proven track record of frustrating personal and inter- personal happiness***. For example, if you think that your mate is angry with you just because she is exceptionally quiet, you are committing the fallacy of *Unsupported Explanation*. By thinking like this, you may well find yourself responding to your mate in ways that negatively affect your relationship. Similarly, if you evaluate your job loss as the worst that could happen, then you are committing the fallacy of *Awfulizing*. By thinking like this, you will have made yourself unnecessarily miserable. In the course of this book these and many other important fallacies, which have historically been investigated by philosophers and psychologists will be discussed and illustrated in a clear and concise fashion, and you will have an opportunity to work on eliminating them from your own active repertoire of ways of thinking. (A word of caution to keep in mind: there is no human being who is 100% free of fallacious thinking; and you are thinking irrationally if you think you can ever be!)

There are five general classes of fallacies that are treated in this book. Each of them is addressed in a separate chapter. They are, respectively, (1) Fallacies of Fact- Gathering, (2) Logical Fallacies, (3) Fallacies of Meaning, (4) Relevance Fallacies, and (5) Fallacies of Making Value Judgments. Each of these general classes include a number of specific fallacies. Although each of these specific fallacies is different, it is sometimes possible for more than one fallacy to describe the same thinking error. So don't be surprised or puzzled if, in applying the fallacies to life situations, two or more of these fallacies seem to fit. It is not uncommon for there to be more than one sensible way of looking at the same thinking error.

Nor is it unusual to find that there are several different thinking errors committed by the same person in the same context. Fallacies often *cluster* about one another. That is, once a person commits one fallacy others will often follow. As you will see, there are also certain fallacy "syndromes" or faulty thinking patterns consisting of several different fallacies in combination. (For example, in Chapter 6 you will be introduced to the "Slippery-Slope/Awfulizing/I-Can't- Stand-It-Itis Syndrome.") So, when you catch one fallacy, don't be surprised if there are more that follow.

SOME TIPS ON FINDING FALLACIES IN YOUR OWN THINKING

Fallacies in thinking do not always glare in your face, especially if *you* are the one committing them. Some special efforts may therefore be helpful in "digging them out" of your thinking. One fruitful way of accomplishing this feat is to ask yourself certain kinds of questions that invite analysis of your thinking. Let's briefly explore three such questions.

(1) Getting At Fallacies By Asking "Then What?"

People often experience anxiety over the things that they irrationally think will happen as a result of other things that have happened or may happen. For instance, suppose you are experiencing anxiety over the prospects of a divorce. What are you telling yourself about what will happen if you do, indeed, get the divorce? "*Then what?*" you may ask yourself. "Well, "I'll never find anyone else." *And, then what?* "My life will be empty and solitary." *And then what?* "I'll probably lose my mind completely." Notice that by asking yourself the "Then what?" question re- peatedly you can get clearer about the things you are telling yourself to get yourself upset. You can bring to light the *inferences* you are making about the disturbing event so that you can better see what is disturbing you about it. Many times, as in the present case, the line of reasoning that you sell yourself is riddled with fallacies. For example, where is your proof that you will never find *anyone* else?

(2) Getting At Fallacies By Philosophically Asking "Why?"

You can also ask yourself for the reasons *why* you feel or think a certain way with an eye toward examining these reasons *philosophically.* Often the reasons that you may have are surprisingly flawed and do not hold up under such *careful* inspection. For example, *why* do I think that I am worthless? "Well, because I can never do anything right." But, is it really true that I can *never* do *anything* right? Don't I at least do *some* things right? So am I really worthless? Another example: Why do I feel so miserable about that mistake I made at work yesterday? "Well, because I feel so incompetent." But can I *reasonably* expect that I *never* goof up? *Must* I be *perfect*? Am I not demanding what is *humanly* impossible of myself? Notice how by philosophically questioning your beliefs and feelings by asking "why?"--questing for rational justification for your beliefs, trying to be objective about your own thinking, carefully examining your terms--you can discover the soft spots in your own thinking.

(3) Getting At Fallacies By Asking "Why is that a reason?"

Sometimes you may have to carry your philosophical inquiry to an even deeper level, looking carefully at the *reasons for your reasons* in order to discover fallacies. That is, once you have given yourself a reason why you are thinking or feeling a certain way, you may have to ask yourself why your reason *is* a reason in the first place. For example, why are you so angry at your spouse? "Because he/she didn't even call to tell me he/she would be home late." But why is that *a good reason* why you should be so angry at him/her? "Well, because he/she didn't even have the decency to call. I can't stand to be treated like that by anyone, especially someone who is supposed to care about me." But *must* people who care about you *always* do what you want them to do? And, if they don't, does that make them *100% indecent*? Can you really *not stand* what your spouse did? Is it rather that you would have *preferred* it had he/she phoned you? Notice how the "Why is that a reason?" question can help you to get beneath your surface layer of reasons, to *further* layers that may contain troublesome irrational ideas.

ABOUT THE PRACTICE SCENARIOS

At the end of each chapter (beginning with Chapter 2 through Chapter 8) you will find a number of practice scenarios dealing with life decisions, interpersonal relations, assertiveness and stress. In each of these, at least one of the fallacies discussed in the chapter has been committed; or, as in the cases of Chapters 7 and 8, some error in assertiveness or the managing of stress has been made. You will be asked to identify the faulty thinking or inappropriate behavior and to consider how it might have been avoided. These practice scenarios have been primarily designed to help you to build a better understanding of the nature and significance of the fallacies as they relate to actual life situations concerning decision making, interpersonal relations, assertiveness and stress management. Indeed, as you work through them, you are likely to note a number of situations and styles of thinking that resemble ones from your own stock of life experiences.

One very effective way of working through these scenarios in a class context is to divide the class into smaller groups, for example, a class of 30 could be divided into six groups with five members in each. Groups are then assigned specific scenarios to analyze. Groups then take turns in reading and analyzing their assigned scenarios before the entire class. In this way, every class member is given an opportunity to participate and reflect upon the problems raised by the scenarios.

SOFTWARE APPLICATIONS

(1) BELIEF-SCAN FOR WINDOWS: Artificial Intelligence for Detecting And Diagnosing Faulty Thinking

This text is supported by a computerized program called *BELIEF-SCAN.* This software option is a form of artificial intelligence that can be used to check your own beliefs or the beliefs of other people for faulty thinking. When you type in your own beliefs or somebody else's, this program "scans" these sentences for possible fallacies. *BELIEF-SCAN* tells you why the thinking in question may be fallacious; it offers suggestions for avoiding such thinking in the future; it provides definitions of the fallacies that have been detected; it concisely summarizes findings, diagnoses thinking problems (using the same non-technical language used in this text); and it issues cautions and recommendations for remedying these problems.

In a survey of 26 college students who used *BELIEF-SCAN* in conjunction with an earlier edition of the present text, most students (23 out of the 26) reported that use of the computer program helped to improve their critical thinking skills. Most (24 out of the 26) also reported that the program provided useful insights into their own thinking about personal matters. Moreover, most (22 out of 26) reported that use of the program helped them to notice fallacies in their thinking that they previously did not *know* they were committing.

The students in this study were asked to write essays reflecting their own thinking about current matters in their lives. Each student wrote essays of about 350 words each week for a 6 week summer class meeting 4 hours per week. The average amount of time spent per week on this assignment was about 1.5 hours.

Each week, the students were required to turn in *BELIEF-SCAN* "Summaries of Possible Fallacies Found" and "Overall Evaluations." These computerized reports displayed the fallacies students committed as well as diagnoses of indicated thinking problems. Since *BELIEF-SCAN* Summaries and Evaluations do not display the actual sentences that were entered, the students' privacy was protected and they were able to get as personal as they wanted without concern that the instructor would later see what they had entered.

Summaries and Evaluations submitted by students were compared with *pre*-test results of another program called the *Practical Logic Assessment Inventory*. This program asks subjects a battery of questions designed to assess their thinking errors, and, on the basis of subjects' responses, generates a list of fallacies frequently committed. This data was in turn used to establish program efficiency ratings. According to this analysis, *one third* of the fallacies detected by *BELIEF-SCAN* (as found by the program in at least two entered sentences) were fallacies with which students did not even perceive themselves to have any problem. This finding suggests that use of *BELIEF-SCAN* in conjunction with this text can be effective in helping *you* find fallacies in your own thinking that you otherwise would not even have known you were prone to commit. It also supports a general theme of this book, namely that people may often have a difficult time in recognizing their *own* faulty thinking.

You should bear in mind, however, that *BELIEF-SCAN* is not intended as a substitute for your own ability to limit your fallacies. While it can find fallacies, it is still up to you to avoid them. In this regard, *BELIEF-SCAN* is like a spell checker. A spell checker can catch misspellings but it does not necessarily make you a good speller. Similarly, you will need to work hard to develop your own thinking skills. Indeed, the computer cannot do your thinking *for* you. Moreover, while *BELIEF-SCAN* is extremely sensitive, it is not perfect. It is itself a fallacy to suppose that because the program has not found a fallacy in a given sentence that therefore there *must* not be any.

(2) Suggested Practical Exercises Using BELIEF-SCAN:

What follows is a list of some suggested exercises using *BELIEF-SCAN*. (See also *Further Practical Exercises* in Chapters 2 through 6.)

1. On a weekly basis in conjunction with your logging (see below section on "Keeping A Personal Log"), write a 350 word essay reflecting your thinking about issues in your own life. These essays may include your reactions to what others have said or done or what you yourself have said or done. Run a *complete BELIEF-SCAN* analysis of your essays. Make hard (as well as soft) copies of *BELIEF-SCAN* Summaries and Evaluations. Maintaining copies of your personal essays for your records, submit all *BELIEF-SCAN* Summaries and Evaluations to your instructor for credit. In addition, examine your summaries for patterns of fallacies that you have repeated more than once. Select and list your "Top Three"

fallacies, and discuss whether you think these fallacies are problems for you and what strategies you might use to try to avoid these fallacies in the future.

2. Write down several statements made by one or more people with whom you are acquainted (friends, relatives, co-workers, etc.) and whose statements you suspect may contain faulty thinking. Disguising the identity of these people (do not enter actual names, etc.) run the statements through a *complete BELIEF-SCAN* analysis. Using these reports as a basis, write a clear and concise evaluation of the person's thinking. (Leave out or disguise any descriptions that might reveal the person's actual identity). Explain what is faulty about the thinking and how you think the person could have more rationally dealt with his or her subject. Submit your written evaluation, together with supporting *BELIEF-SCAN* Summaries and Evaluations, to your instructor.

3. Write an essay of approximately 500 words in length on a controversial issue of your choice (for example, abortion, gun control, mercy killing, etc.). Take a stand on the issue, arguing for a particular conclusion and supporting your argument with what you think are good reasons. Check your thinking for fallacies by running your essay through a *complete BELIEF-SCAN* analysis.

4. Pick out an essay or editorial written on a controversial issue (for example, abortion, gun control, mercy killing, etc.). You may use newspapers, magazines, journals, anthologies containing essays on controversial moral problems, etc. Check the essay for fallacies by running the essay through a *complete BELIEF-SCAN* analysis. Using your *BELIEF-SCAN* reports, write a critique of the essay. Describe any weak spots (fallacies committed) in the author's thinking and/or any of its strong points (well supported arguments, fallacies avoided, etc.). Suggest and provide support for any constructive changes in the author's thinking.

THINKING ABOUT YOUR OWN THINKING: KEEPING A PERSONAL LOG

It is strongly recommended that you keep a daily log of self-observations concerning the ways in which fallacies have related to your personal life. By keeping such a log, you will give yourself an opportunity to reflect upon your own thinking and, thereby, to gain greater insight into the weaknesses as well as the strengths of your own thought habits. In general, students who have kept logs have reported greater improvements in their rational thinking skills.

You can begin your log just as soon as you have done the reading in Chapter 2. Here are some suggestions about information to include in your log:

1. Mention the date (month/day/year) on which you committed a given fallacy. This can help you to track your progress later on.
2. Mention, by name, the fallacy you have committed.
3. Explain the circumstances under which you committed this fallacy.
4. Say how you might have avoided this fallacy. That is, mention other things you could have thought, said or done that would not have involved your committing a fallacy.
5. Mention some *behavioral goals* (things you intend to do in the future) that you think might help you to avoid this fallacy in the future.
6. Mention any cases in which you stopped yourself from committing a fallacy and instead responded appropriately.

   Keep in mind that as you become more skillful at avoiding fallacies, you will find situations in which you begin to think irrationally but quickly realize it, stop it, and redirect your thinking along more rational lines. This is indeed a sign of genuine progress.

7. Describe any instances in which you could have better asserted yourself, indicating how you could have done so. (See Chapter 7.)
8. Describe any specific, future plans you have made for practicing being more assertive. (See Chapter 7.)
9. Describe any stressful experiences, indicating any ways in which you might have been contributing to that stress. (See Chapter 8.)
10. Mention any ways in which you might have reduced your stress. (See Chapter 8.)
11. Describe any specific, future plans you have for more effectively managing your stress. (See Chapter 8.)
12. Mention anything else you think is significant.

How long should you continue to log? While some users of this book have preferred to continue logging several months after completing all chapters, it is recommended that you continue to keep a log for at least two or three weeks after you have completed your readings. In any case, keep your book handy. You are likely to find reason to consult it long after you have completed the reading.

If you are using *BELIEF-SCAN* in conjunction with this text, you may find it helpful to use computer analyses of your thinking to generate data for your log. You may also use *BELIEF-SCAN* to analyze the beliefs you enter into your log. Here are a few representative remarks from students who have used *BELIEF-SCAN* in connection with logging:

"I was able to record the entries and all responses every time I used BELIEF-SCAN so my log was easy to maintain"

"My personal log gave me ideas...to enter into BELIEF-SCAN."
"In doing a personal log, I would edit it through the program. I was able to make myself clearer, and use fewer words to explain

myself, be more to the point, and more tactful."

"In my log I put the full scenarios, and then used BELIEF-SCAN to verify certain fallacies in my statements."

IMPROVING YOUR THINKING SKILLS: SOME PRELIMINARY CONSIDERATIONS

Keep in mind that changing your thinking habits takes practice and effort on your part; so don't expect some magical change to happen overnight with little effort. Here is a list of things to consider and to work at if constructive change is to occur:

(1) Carefully working through the readings and practical exercises in this text, develop a good understanding of each of the fallacies, learning their names and definitions so that you are able to quickly identify them as they occur in actual life contexts. The fallacies that will be discussed in this book may be "easy to see" once they have been pointed out to us by somebody else. Moreover, many of us may often have an unclear sense that there is "something wrong" with a faulty line of thinking even before the fallacy in question is pointed out. Nevertheless, having a classification system that allows you to place faulty thinking under specific names and definitions can, be very useful for purposes of noticing these fallacies on your own *without* their first having to be identified *for you*. Moreover, such a classification system makes it easier to see *precisely and clearly* just what is wrong with a faulty line of thinking.

(2) Instead of saving this understanding strictly for the classroom, make a concerted effort to *apply* what you have learned to actual life situations *outside* the classroom.

(3) Work at developing a critical reflective attitude toward your own thinking and reasoning. This means making an ongoing, special effort to assess your *own* thinking by asking yourself if it commits any of the fallacies you have studied; and, if you determine that it does, by asking yourself what a more rational way of thinking would be. This may sound easier than it actually is.

Most people tend to think about things other than their own thinking. Thinking *about* your own thinking (where the subject of your thinking is *itself* your thinking) is an activity that may take some getting used to for many of us. Using *BELIEF-SCAN* to analyze your own thinking may also be helpful in this regard.

(4) Keep a detailed daily log, which allows you to record as well as to further examine and assess your own thinking.

(5) Work at it *behaviorally* by acting in ways that are consistent with your newly formed rational beliefs. Don't merely pay lip service to them while doing the same old self-defeating things.

(6) Keep at it. It is easy to become lax in your efforts, and to drift back into a routine of faulty thinking. Have patience with yourself.

(7) Don't expect to do a *perfect* job in spotting all the fallacies you might commit. As will be stressed in Chapter 6, if you demand perfection, you are bound to cause yourself unnecessary frustration. While it is reasonable to work on improving your thinking skills, it is irrational to expect perfection.

(8) Keep in mind that the more you practice at applying the fallacies to your life, the more adept you will get at it. So, if you feel a bit clumsy at first, don't worry. Just keep at it.

(9) At the beginning, some people sometimes become extremely zealous in looking for fallacies. When this fallacy hunt is directed toward other people, who may not be as aware of the value of giving up their faulty thinking as you are, these others may become resentful of your attempt to change or correct them. This does not mean that you should not tell others when they are committing fallacies. Indeed, doing just that can be an important way of stimulating change among others with whom you share intimate relationships (for instance, your spouse, mate, close friend or parent). But, if you do, don't be surprised if you encounter some resistance. It is usually good policy to use discretion, tact and diplomacy when criticizing others. In any event, it is important not to allow yourself to be discouraged by others who may ridicule what you are trying to do. In many cases, those who offer such

resistance themselves later nicely "come around" to the new way of thinking.

(10) If you are aiming at improving intimate relationships such as those with a spouse or mate, don't be shocked if things get bumpier before they get smoother. One common reason for this is that, as you change, others around you may not themselves know how to deal with these changes. Your old ways may also be ways of life for them too, so much so that they may lose direction when you surprise them with new responses to old stimuli. For example, suppose you used to yell, scream and call your spouse nasty things whenever he/she criticized you; but now suppose that you respond by rationally discussing these criticisms with him/her. In this case, your spouse may be ready to respond in his/her usual way (for example, screaming and yelling back at you or withdrawing and acting like a victim). But now he/she can't get his/her usual cue from you. Indeed, he/she may become quite frustrated at this, going to some lengths to get a rise out of you. However, if you keep working at making rational responses, your spouse may also adjust to your new behavior and positive change in your relation can then take place. So it will be important for you to have patience with your significant other in such cases.

(11) You should not assume that your relation is *guaranteed* to improve even if you do a good job at avoiding fallacies in your thinking. There may be problems with your relation that could benefit from the help of a trained professional, for example, a marriage counselor. Moreover, you are not the only person in your relation who might need to do some changing. Where your significant other(s) is (are) willing, it is also a good idea for each of you to share in the use of this book. In any event, since many relational problems are rooted in bad thinking, making positive changes in the way you think is likely to make important contributions to happier, healthier relations.

(12) Finally, one concern that some people have about being logical about life is that they will lose their power to have emotions, thereby turning into some kind of unfeeling robot. This view is entirely unfounded, however. This is because there can be

rational emotions as well as irrational ones. For example, it can be rational to get angry or disturbed at a wrong done to you and to take constructive measures to rectify it. However, it would be irrational to become enraged (to "fly off the handle") and to speak and act in ways that defeat your own purposes. While irrational emotions normally thwart your ability to be happy, rational emotions are more likely to motivate you toward greater happiness. One main purpose of this book is to help you learn how to avoid the irrational emotions, not to teach you how to be a robot devoid of all emotion.

BENEFITS OF AVOIDING FAULTY THINKING: SOME TESTIMONY

Below are some representative remarks made by students who have used earlier versions of this book. These remarks may help you to gain a better idea of the sort of benefits that might accrue by learning how to avoid faulty thinking in your practical existence:

". . . Being able to avoid or correct fallacious thinking has helped in my relationships with others and in how I feel about myself. I believe my decisions are more sound. I know that when I express my thoughts or feelings to others I am much clearer than I used to be and so avoid confusion and unnecessary or unfounded bad feelings. The information has also helped because I can see when others are committing fallacies. I realize that I need to ask for clarification sometimes. I also know that what others say is not always so. This has made me less dependent on what others think and feel, especially about me."

"I feel more relaxed about certain situations. I am working at not getting riled up (internally) about situations that are not that important. Also I am establishing my priorities about what is important and what isn't."

"The application of this material has certainly made a difference in my life in that I now have a start, an anchor to which change in my emotional response to life can be tied. My thinking has changed yet continued practice will be necessary. My feelings have also changed but to a lesser degree."

"Slippery Slope, Had Things Been Different, Fallacy of the Whole, Parroting, Bandwagon, Misuse of Pity, The World Revolves Around Me, and Awfulizing are fallacies I seem to commit most frequently. I have already noticed in my behavior and communication that I am sometimes aware of these now before I begin to commit them, or even during committing them. ..."

"In a recent labor union negotiation, where I represented the department head, I was very amazed at the number of fallacies that were committed. When you realize the fallacies that others use, you can more easily spot the weakness in their arguments which gives your position more validity. . . ."

"I am a very demanding individual, especially on myself. I guess you might say I'm my own worst enemy. I always wanted things done right and expected perfection in myself. . . . Since I have come to understand fallacies, I now realize that my thinking was wrong--and at times my faulty thinking did affect my happiness. Now I realize it was me who did it to myself. I now watch myself, do my job to the best of my abilities and do not demand as much of myself and of others. My whole feelings have changed--now I can laugh at things I never would have laughed at before . . . I still commit fallacies but now I know that when this happens I can change my way of thinking."

"At first I recognized the fallacies others were committing. It is harder to take a good honest look at myself. I am, though, and I am trying mostly to make changes in my assertiveness skills. . ."

Keeping in mind the points made in this chapter, let's now get started on the task of learning how to identify and avoid faulty thinking that can be harmful to your happiness.

CHAPTER TWO
FALLACIES OF
FACT GATHERING

WHAT ARE FALLACIES OF FACT GATHERING?

Fallacies of Fact Gathering are mistakes involving distortions of *fact or probability*. Since factual and probab- ility distortions can happen in a number of different ways, there are a number of different kinds of Fallacies of Fact Gathering. The following is a discussion of each.

1. THE STRAW MAN

This fallacy occurs when you present a claim or issue in such a way that it becomes easier to "blow over" like a straw man. The *motive* (whether or not you are aware of it) you might have for doing this is to make your "side of the story" sound more convincing by making "the other side of the story" harder to "swallow."

Words like "always" and "never" are sometimes cues that you are falling into this fallacy. For example, Tom says, "My wife never cleans the house" where the truth is that Tom's wife does not clean the house as much as Tom would like her to; or "She's always "complaining." It is unlikely that she's *always* complaining!

The use of strong emotional language is also sometimes a cue that this fallacy is occurring. For example, John says "My father *exploded* at me when I asked him for the car keys," while the actual case was much less dramatic.

2. SLIPPERY SLOPE

This fallacy is a kind of Straw Man in which you exaggerate the negative *effects* of an action or event so that you make them seem harder to "swallow." For example, Jane says "If my husband leaves me, then I will be left totally helpless and destitute"; and Sarah says "If my daughter marries that boy, then she will move away and I will never see her again." As in the case of the general straw man, this fallacy may involve the use of strong emotional language. For example, notice that Jane uses the words "helpless" and "destitute." Similarly, notice Sarah uses the words "totally" and "never."

In some cases of Slippery Slope, you exaggerate the negative effects of an act or event by suggesting that these effects will "snowball," so that one bad effect will gradually lead to further even worse ones. For example, Mort, a college student, tells himself that "If I don't get to bed on time tonight, I won't be able to think tomorrow; I'll mess up on that big exam; I'll get a lousy grade in that course; my grade point average will drop; my grade transcript will look like hell; then, when I apply for a job, no one will ever hire me"; and I'll never amount to anything!" Even if Mort does not get as much sleep as he would like, that does not mean that he will not be able *to think* tomorrow. Nor does it mean that, even if he does poorly on the exam, *no one* will *ever* hire him or that he will "*never* amount to *anything*." Mort could have avoided this slide down the Slippery Slope by noting the strong language by which he, step by step, exaggerated the results of getting to bed at a certain hour. (Note that in Chapter 8, the role of the "Then what?" question will also be discussed in managing Slippery Slopes in anxiety and other emotions.)

3. MAGNIFYING RISKS

In this fallacy the *probability* of something bad happening gets blown out of proportion. For example, because her physician told Jennifer that there is a 1% risk that the surgery she needs can leave her without motion in her legs, Jennifer says, "Now I'll probably end up in a wheel chair for the rest of my life." And since he once got a speeding ticket when he took Highway 1 to work, Ron refuses to ever take that route again because, he says, "I'll probably end up getting another ticket no matter how fast I go."

In addition to the word "probably," examples of other words that are sometimes used to magnify risks are "likely," "risky," "definitely," "certain," "sure," and "know." This does not mean, however, that whenever these or related words are used a fallacy has been committed. It is committed only when you use them in contexts where the evidence clearly does not support their use. For example, Paul is afraid to fly in an airplane because he says he "knows" and is "sure" that the plane will "definitely" crash if he flies in one. However, relative to the actual track records of most major airlines, such a risk claim would be highly exaggerated.

While the Slippery Slope typically involves Magnifying Risks, these two fallacies are not the same. The major difference lies in the fact that Magnifying Risks always involves the exaggeration of the *probability* of an event happening rather than the exaggeration of *what* will or might happen. For example, the inference from "The weather report said that there is a slight possibility of afternoon showers" to "Now it's probably going to rain this afternoon" merely exaggerates the *probability* of its raining this afternoon. On the other hand, the inference from "The weather report predicted late afternoon showers" to "Now it's probably going to rain *all day*" also exaggerates *what* will probably happen. Unlike the first, this second inference therefore also commits the Slippery Slope fallacy.

4. HALF TRUTH

The fallacy of Half Truth occurs when you state some parts of a situation and leave out others so that you create a false understanding of the actual case. In common cases of this fallacy (sometimes called "card stacking"), you "stack" things in your own favor. For example, Clio says, "My boyfriend didn't show up for our date, and he didn't even bother to call," leaving out, however, the important fact that her boyfriend was in an automobile accident on his way to her house. This might in turn lead Clio to unfairly criticize her boyfriend. For example, she might unfairly call him "unreliable" and "irresponsible."

5. THE "HAD THINGS BEEN DIFFERENT" FALLACY

This fallacy arises when you speculate about what would have happened had something else been the case. For example, Nancy says, "If I had tried harder to defend myself then my father would not have been able to sexually abuse me." Since this statement is about the results of an event that never actually happened ("I tried harder to defend myself"), it cannot really be proven. Indeed, it assumes that Nancy's father would not also have tried harder, which is something Nancy could not know. Nancy's statement could be made more reasonable if the word "maybe" (or "possibly" or "perhaps" or the like) is added. She could (reasonably) say, "If I had tried harder to defend myself, then *maybe* my father would not have been able to sexually abuse me." But notice that this statement is much less definite about what would have happened, had she tried harder.

As the above example suggests, you can use the "Had Things Been Different" Fallacy to make yourself upset. For example, Nancy could tell herself that it is her own fault that her father sexually abused her. In order not to fall into this trap, you should be aware of the difficulty of proving things that never really happened.

The "Had Things Been Different" Fallacy can also be used to tell yourself "stories." For example, "If my wife hadn't taken my mind off of driving, then I wouldn't have gotten into that automobile accident." Well *maybe*, but then again maybe not.

6. FALSE COMPARISON

A false comparison occurs when you expect two (or more) things to be alike in a certain respect just because they are alike in some other respect. For example, if Jim expects his present wife to divorce him just because his ex-wife did, then he is making a false comparison. In this case Jim is probably overlooking important differences between the woman in question, as well as between the circumstances of his two marriages. (Jim himself may have changed, his financial conditions may be different, etc.) In any event, Jim has certainly not pointed to any similarities between both women, which would justify his expectation of divorce. The fact that each woman is, or has been, his wife is not, in itself, a similarity, which would justify this expectation.

In contrast, a true comparison is one where there exists enough similarities and no important differences between the things being compared. For example, if you buy two pairs of slacks, one black and the other brown, and if each is the same size and the same brand, then if the black pair fits, it is likely (but still not certain) that the brown pair will fit also.

7. UNSUPPORTED EXPLANATION

An unsupported explanation is one that does not adequately explain all the facts that need explaining. For example, Janis says "My husband is having an affair with another woman" in order to explain her husband's coming home late the past week. This explanation could explain why there are lipstick stains on his collar. However, suppose that the following are also facts: the color of the lipstick stain matches the color of Janis' own lipstick; her husband's boss backs up her husband's claim to have been working late the past week; her husband had been in his office during each of the nights in question to receive telephone calls from Janis. Given these facts, the explan- ation in question would be an unsupported one.

There may be several different explanations that will explain a situation. Moreover, there may be no more reason to prefer one of these explanations over the others. Where this is the case, you can commit the fallacy of Unsupported Explanation by "buying" one of these explanations and rejecting the others. For example, Lynn is waiting for her husband Myer to come home from work but he is about an hour late. In this case, Lynn would commit the fallacy in question by jumping to the conclusion that Myer has gotten into an automobile accident. While this is possible, it is at the time no more supportable than a number of other possible explanations, for example, that Myer stopped off to get something at the store, that he was caught in traffic, that he worked later than usual, and so forth. In such a case, where there are several equally supportable explanations, it is rational for you to withhold your judgment until you have collected more facts. For example, by phoning Myer's place of work, Lynn can check out the "working late" explanation or she can simply wait a bit longer to see if he comes home. Given that several explanations are equally supportable at a given time, it is rational to do the easiest things first. For example, calling Myer's work is a lot easier than combing the streets for a wrecked car. However, until the facts support the "auto accident" explanation over any of the others, it would be irrational for Lynn to accept this explanation.

Perhaps the "auto accident" explanation is the one that Lynn fears the most. However, don't be moved by your fears into focusing your attention on the worst explanations when other possible explanations are equally as supportable. The fact that you are afraid that an explanation might be true is hardly a good reason for thinking that it is true. By bearing this in mind, you are likely to find it easier to keep your composure and to avoid self-defeating behavior. For example, Lynn's failure to keep this in mind could lead her to frantically ride up and down the city streets looking for Myer's wrecked car while Myer sits at home wondering where Lynn is!

8. OVERSIMPLIFYING

This fallacy occurs when you make a more complex situation out to be simpler than it really is. For example, Arthur says that, "All I have to do to improve my grades in school is to study." This would oversimplify things because effective study involves a number of factors including how long Arthur studies, what he studies, and how he studies. Some English phrases that suggest oversimplifying are "all that is necessary," "in a nutshell," "the only thing," and similar ones. In basing your decisions on oversimplified beliefs about things, you are apt to act in self-defeating ways. For example, Arthur may find that even though he did, indeed, study, he still received failing grades.

9. HASTY GENERALIZING

Hasty Generalizing occurs when you "jump to conclusions" about all or most of a group. For example, Iris concludes that "Most men are child sexual abusers" just because her father sexually abused her and her husband sexually abused her daughter. This is a "hasty" conclusion because Iris is basing her belief about almost all men on just two men. However, even if Iris based her belief on a lot of men she knows, her conclusion could still turn out to be "hasty." For example, if she based it on only men who were themselves sexually abused, then she would still be "jumping to conclusions" about almost all men. This is because even if many men who were themselves sexually abused are child sexual abusers, this may not be true of many men who have not themselves been sexually abused.

It is not uncommon for people to dismiss any attempt to generalize about a group by saying, "Ah, you're just generalizing," as if *all* generalizations were always defective. However, the generalization that "All generaliz- ations are no good" is itself a hasty generalization. While some, indeed many, generalizations may be defective, there are also some that aren't. For example, highly confirmed laws of nature such as the "universal law" of gravity are not defective.

To be acceptable, your generalization should avoid three related pitfalls. First, you should make draw on the basis of a *sufficient number* of group members. In ordinary life situations generalizations about groups having thousands, millions or even billions of members may be drawn from only a few cases ("All men are child molesters [because my father molested me]," "All Fords are `lemons' [because I once had a Ford `lemon'"]) While the term "sufficient number" admittedly lacks precision, it is clear that observations about only a few members of such large groups will not alone adequately support generalizations about all or most members of these groups.

Second, your generalization should be based not only on a sufficient number of group members; it should also be drawn from group members that are *sufficiently different.* For example, as stated, the generalization, "Most men are child sexual abusers," as derived from observations of many men who were themselves sexually abused, would also be hastily drawn. To avoid Hasty Generalizing, this gener- alization would need to be drawn from a sample that also included many men who were *not* themselves sexually abused.

Third, your generalization should avoid *inconsistency.* For example, Jack claims that *everybody* in his school dislikes a certain teacher, Mrs. Applegate. Yet he is aware that Jamie, a schoolmate, is very fond of this same teacher. In this case, Jack might consider "weakening" his claim to "Most people in my school dislike Mrs. Applegate" provided however that this weaker generalization is supportable.

10. STEREOTYPING

Stereotyping occurs when you "lump" certain people together without paying attention to their individual differences. For example, when Sue says that "All men are after just one thing" she is stereotyping all men. Stereotyping often uses strong emotional words. For example, George thinks that "All blonde women are dumb." Notice that he uses the word "dumb." Stereotyping may also employ *positive* emotional language as in "All blonde women are sexy." While being regarded as "sexy" may, in some respects, be a desirable trait, such a characterization of all blonde women would still be a stereotype.

Stereotyping may result from Hasty Generalizing (see above). For example, Charles thinks that all Mexicans are thieves because he once employed a Mexican who stole from him. However, while Hasty Generalizing can lead to stereotyping, stereotyping can also lead to Hasty Generalizing. For example, Bob's parents have taught him to stay away from gay men because "Most homosexuals are child molesters." Thus when Bob reads about a gay man who is charged with child molestation, he hastily reaffirms his old view that most gay men are child molesters. In such a case, the fallacy of Hasty Generalizing arises *after*, and as a consequence of, the Stereotyping!

Note that Stereotyping may employ the term "most" as well as "all." Thus, when Jim, who thinks that "All Blacks are criminals," gets to know Al, a Black who is not a criminal, he proclaims that "Al is not like the rest of them. He's just an exception. Most of them are still criminals." In such cases evidence in favor of giving up the stereotype is instead held to be consistent with it.

All Stereotyping involves *prejudgment*. That is, persons are judged negatively or positively without adequate evidence. Consequently, Stereotyping prevents a realistic assessment of individuals. Since satisfactory interpersonal relations require relating to one another *as individuals*--as distinct persons with unique personal qualities, it is easy to see how Stereotyping can and often does seriously impede interpersonal relations. For example, since Frank perceives "beautiful women" as "sex objects" and as "pretty play things," he fails to see these women as distinct persons with unique personal qualities. As a result, he is unsuccessful in establishing lasting and mutually gratifying relations with the women he dates.

11. IGNORING THE PAST

When you predict something, you are making a judgment about the future. Since the future has not yet happened, this judgment must be based upon information about what has already happened in similar cases. You commit the fallacy of Ignoring the Past when you ignore the past in making a judgment about the future. For example, Aimy commits this fallacy when she judges that her husband, Art, who is a heavy drinker, will give up the booze. "Art," she says, "has given me his word that he will quit. This time it is different, he said." However, Art has a long past history of promising unsuccessfully to giving up drinking. In this case, if Aimy ignores the past, she is not apt to do something to change the future. Unless she realizes her husband's established drinking pattern, she will, in all likelihood, enable it to continue in the future. By realizing it, however, she could take important steps (for example, ceasing to take the role of enabler, attending a support group, encouraging Art to get help, and so forth), which could affect future outcomes.

12. INSISTING ON THE PAST

While ignoring the past can result in a failure to make positive changes in the future, it is also possible to misuse information about the past. In particular, just because something has regularly happened in the past does not mean that it must, necessarily, continue in the future. For example, Kirk commits this fallacy when he judges that, since he and his wife have had a "rocky" relationship in the past, it must always be that way in the future and that, therefore, he might as well forget about even trying to change things. Indeed, it is possible to correct past mistakes in the future so that the future does not have to mirror the past. For example, Kirk and his wife could see a marriage counselor. But by insisting that the future must be like the past, Kirk has lost an important opportunity to do something constructive about the future of his marriage.

13. ASSUMING THE CAUSE

This fallacy occurs when you believe that one event is the cause of another event simply because the one event happened before the other. For example, Jerry, who has recently lost his job, believes that "Since I was let go shortly after they hired that new guy, that new guy must have had something to do with it." In this case there may really have been a connection between Jerry's getting fired and the new guy being hired--the management may have hired the new guy to fill Jerry's position; however, Jerry needs more information in order to come to this conclusion; for it may have just been a coincidence. Just because one event (Jerry's being fired) comes after another event (the new guy being hired) does not mean that the two events are connected.

The fallacy of Assuming the Cause can lead you into rituals, superstitions, and unnecessary restrictions on your activities. For example, Harry refuses to wear an expensive new suit to work because the last time he wore "that damn thing" his car broke down and he was mugged.

14. QUESTIONABLE CLAIM

A Questionable Claim is a claim that you make that cannot be (reasonably) accepted unless you are able to give further evidence to back it up.

In some cases, this fallacy amounts to having no evidence whatsoever for your claim. For example, Jack, an elderly person, thinks that his adult children don't really want him to live with them. His belief, however, is not something that he is able to reasonably conclude from anything his children have actually said or done. In fact, Jack may be allowing his fear of rejection to get the better of him.

In other cases, this fallacy arises when you make a claim that is of such a nature that it cannot even in principle be proven or disproved. If someone told you that the reason why your car won't start is that there are invisible, undetectable gremlins in your gas tank, you would not be able to disprove the claim; nor, however, would you be able to *prove* it! Similarly, claims about the human *subconscious* may be of this kind. How, for example, could you get down deep into your own subconscious to disprove the charge that anything you do, even helping others, is really motivated by some "hidden" desire for sexual gratification. Thus, Amy tells Bob that he only *thinks* he cares about her; but what he really cares about is his own desire to "have his way" with her. Since, by definition, what is in his subconscious is outside his own reach (as long as it *remains* "tucked away" in his subconscious), he cannot even begin to disprove her claim.

In still other cases, this fallacy arises when you hold a belief in the form of a *mindset.* A mindset amounts to a *strong tendency to persist in holding a belief even when there is convincing evidence to the contrary.* For example, John believes that men are better than women in mathematics and will not change his mind regardless of how many examples to the contrary he is shown. (Notice how *stereotypes* such as this one are also mindset-driven questionable claims.) Similarly, Sue persists in believing that she does not have a drinking problem regardless of how much evidence there is to the contrary.

15. THE SMOKE SCREEN

This fallacy occurs when you attempt to defend or justify your claim by appealing to a "smoke screen" of emotional words instead of to the facts. For example, Hector claims that he and his wife are not getting along with each other because she's a "real bitch." Since the emotional term "real bitch" is not itself evidence for anything, Hector needs to be specific about what disturbs him about his wife's conduct; he needs to "stick to the facts." For example, when Hector's counselor asks him what he means when he calls his wife a "real bitch," Hector responds, "She's always complaining to me about something." "Is she really always complaining, Hector?" asks the counselor. "Well a lot of times!" says Hector. With the smoke screen removed, the facts are more clearly seen!

Words used to rate or assess things such as "good," "bad," "reasonable," "unreasonable," "right," and "wrong" can also be used as smoke screens that hide the real facts, or the lack of them. For example, "You should do what I tell you because it's the right thing to do" or "I agree with John because what he says is reasonable" or "You should hire him because he's the best man for the job." However, if something is the "right thing to do" then there must be some fact about it that makes it the right thing to do (for instance, it would prevent pain or injury); and if what John says is "reasonable," then there must be some reason why it is; and if someone is "the best man for the job" then he must have some specific qualifications. It is therefore improper to use such words to defend your claims unless you can also provide the facts.

REVIEW DEFINITIONS: FALLACIES OF FACT GATHERING

1. **STRAW MAN** = *Df.* Distorting what another person has done or said in such a way that it becomes easier to criticize or reject it.

2. **SLIPPERY SLOPE** =*Df.* Exaggerating the negative (bad) effects of an action or event so that they appear hard to accept.

3. **MAGNIFYING RISKS** =*Df.* Turning smaller probabilities of

something bad happening into much larger ones.

4. **HALF TRUTH** =*Df.* Misleading another person about some matter by presenting only some of the facts and omitting other important facts.

5. **THE "HAD THINGS BEEN DIFFERENT" FALLACY** = *Df.* Making unprovable claims about what would have been true had something else been true.

6. **FALSE COMPARISON** =*Df.* Expecting two individuals (objects, persons, etc.) to resemble each other in a certain way solely because they have something else in common.

7. **UNSUPPORTED EXPLANATION** =*Df.* Trying to account for *why* something is true in a way that is not adequately backed up by the facts.

8. **OVERSIMPLIFYING** =*Df.* Making a more complicated situation out to be much less complicated than it actually is.

9. **HASTY GENERALIZING** =*Df.* Jumping to a conclusion about all or most of a group on the basis of an insufficient sampling of group members

.

10. **STEREOTYPING** =*Df.* Using strong emotional language to rate entire groups of people without considering individual differences among group members.

11. **IGNORING THE PAST** =*Df.* Making predictions about what will happen in the future without considering what has happened in the past.

12. **INSISTING ON THE PAST** =*Df.* Thinking that because something was true in the past that it must continue to be true in the future.

13. **ASSUMING THE CAUSE** =*Df.* Assuming that, just because one thing happened right after another thing happened, the first thing brought about the second thing.

14. **QUESTIONABLE CLAIM** =*Df.* Having little or no evidence to back up your claim.

15. **THE SMOKE SCREEN** =*Df.* Trying to justify a claim by substituting emotional language for the facts.

PRACTICE SCENARIOS FOR CHAPTER TWO

EACH OF THE CASES PRESENTED BELOW CONTAIN AT LEAST ONE OF THE FALLACIES DISCUSSED IN THIS CHAPTER. FOR EACH CASE, NAME AT LEAST ONE FALLACY IT CONTAINS AND EXPLAIN HOW THE PERSON WHO COMMITTED THE FALLACY MIGHT HAVE AVOIDED IT. THAT IS, EXPLAIN HOW THIS PERSON COULD HAVE MORE EFFECTIVELY DEALT WITH HIS OR HER SITUATION. (ANSWERS TO STARRED EXERCISES MAY BE FOUND IN APPENDIX B AT THE BACK OF THIS TEXT.)

1. Sally has just met a man named Robert at a party. Since the band is loud, she has not yet had a chance to learn much about him. Here is what she tells her girlfriend while in the ladies' room at the party:

Sally: "Robert seems like a nice guy and all but I really think he pushes drugs. First of all, he drives a new Mercedes. And the way he dresses! Where else would he get that kind of money! I don't think I'll say yes if he asks me out."

2. "I know that he promised me before that he wouldn't have any

more affairs, but this time I know he's serious."

3. Mrs. Tomasoni is a high school Spanish teacher. Although she has earned a reputation for being difficult, she actually gives more "A's" and "B's" than most other teachers in the school. She gives nightly homework assignments that take an average of about one hour to complete. While Mrs. Tomasoni usually grades and returns exams within a few days, this year she took two weeks to grade and return the first exam. The following is a statement one of her students makes to another student about Mrs. Tomasoni:

Student: "My Spanish teacher is the hardest teacher. She never gives anyone an "A" and she piles on the homework. I don't have any time for anything else because I'm always doing Spanish. Not only that, she never gives back the tests. We had to wait almost the whole year just to get back the first test."

*4. Janie's physician has diagnosed her as having mild mitral valve prolapse, a condition of the mitral valve in the heart. Her physician has told her that there is no significant risk that this condition will shorten her life or that she will ever need heart surgery. Here is what she tells her husband:

Janie: "Now I'll probably end up needing heart surgery. For sure, I'll be dead by the time I'm forty!"

5. Mary is in a state of depression over the recent death of her husband who suffered a sudden heart attack when Mary was out shopping. Here is what she keeps telling herself:

Mary: "If only I was home, I know I could have done something to save him."

6. The following exchange occurs between Mr. Crust, a high school physical education teacher, and Jon, one of his freshman students.

Jon: "Mr. Crust, I have a note from the doctor excusing me from taking PE for the rest of this week. I'm not supposed to put too much pressure on that leg of mine. I really hope this doesn't affect my grade."

Mr. Crust: "Look son, I am not going to hold your hand, kiss you on the cheek, and tell you not to worry. This is not nursery school and I'm not your mother. All of you freshman are so babyish and immature."

7. "My son would never take drugs, so don't even suggest it. I don't want to hear another word about it."

8. Lauren is a 21-year-old woman looking for a permanent relationship with a man. She has recently met a man whom she thinks might be "Mr. Right." Here is her reasoning:

Lauren: "My sister married an engineer and they're really happy together. My new boyfriend, Harvey, is also studying to be an engineer. He is two years older than I am, which is the same age difference as between my sister and her hubby. I know I only dated Harvey a few times but I really think it's going to work out between us."

9. Here is a discussion between Flossy and her friend, Sam:

Flossy: "I've decided to quit my job at the factory."

Sam: "Why? What's wrong?"

Flossy: "They just hired some Puerto Rican and made him my supervisor. I just don't trust those people. I figure I better quit before he rapes me or something."

10. "Live together out of wedlock? I never heard of anything more immoral and disgusting. The thought of any daughter of mine stooping so low as to live like some sort of cheap tramp

makes me want to puke."

11. Lester suspects that his wife is having an affair. Here is a discussion between him and his friend, Larry:

Lester: "I think my wife is doin it with another guy."
Larry: "Are you kidding me, man. With who? How'd you find out?"
Lester: "I don't know who. I just have a feelin that she's messin around. I didn't catch her or nothin; but I just got this feelin that somethin ain't right. Ya know what I mean?"

12. "If it weren't for those homosexuals, there wouldn't have been an AIDS epidemic. They're the ones who started it; and now they're gonna be responsible for the destruction of the entire human race."

*13. Maxine is a somewhat overweight, 36 year old housewife and mother. Married and pregnant at the age of 19, Maxine has been a housewife and mother for the past 17 years. She has, however, recently interviewed for a position as a receptionist in a physician's office. When she was notified that someone else had been hired for the job, here is a discussion she had with her husband, Fred:

Maxine: "I know exactly why I didn't get the job."
Fred: "Why Max?"
Maxine: "Because they wanted someone more attractive. I'm too heavy. That's why I didn't get it."

14. John and Rose have been married for only two months. So far they have had difficulty relating to one another. They frequently argue and things seem to be getting worse. When John comes to his father for advice, here is what his father says:

John's Father: "Rose is a woman. Women like to be told things like "I love you." Try telling her that a few times a day and everything will be just fine. That's all you need to do to have a good marriage."

15. Here is a remark that Mr. Sandeli, a high school history teacher of twenty years makes to his wife about his teaching career:

Mr. Sandeli: "I often wonder whether anything I have ever taught those kids has made any important, lasting difference in their lives. Relatively few of them have even come back and thanked me for helping them. I suspect that I have wasted my life doing this for a living.

16. Jackie's husband, Bob, has physically abused her for the three-year period in which they have been married. Here is what she tells a friend about Bob's behavior:

Jackie: "Bob is really a good man. This time he's really sorry about what he's done. I don't think that he'll beat me again."

17. Here is a discussion between two teachers, Mr. Rogers and Mrs. Testa:

Mr. Rogers: "Do you have James Rio in your class this year?"
Mrs. Testa: "Yes, I do, as a matter of fact."
Mr. Rogers: "How's he doing? He got a C from me last year."
Mrs. Testa: "He also got a "C" on my first test. Obviously a "C" student!"
Mr. Rogers: "Yes, indeed he is."

*18. The following is between two girlfriends, Polly and Patty:

Polly: "My husband is a real Sweet-Old-Boy; you know, S.O.B."
Patty: "Why? What did he do this time?"
Polly: "Oh, he acts like such a bag of horse manure; you know,

a real turkey."

Patty: "Yeah, I think I read you loud and clear, Pol. My little darlin acts just like that too."

19. "She's the right one for me. If I can't have her, then nothing else in my life can ever matter."

20. Here is some advice that Pete, a white man, gives to his 12-year-old son:

Pete: "The man who shot and killed my best friend was black; and the man who mugged me was black too. I wouldn't trust a black as far as you can throw one. They're all no good."

21. Rudy and family are planning their yearly vacation. His wife and kids want to go to Disney World. Here's what Rudy tells his wife:

Rudy: "Honey, I don't think we should go to Disney World for our vacation this year. Do you remember what happened last time we went there? Remember? The car broke down and the kids got sick too. Can't we go somewhere else less risky?"

22. History Professor (addressing students): "You are permitted three unexcused absences. If you exceed this limit and cannot provide any legitimate reason for being out, then your final grade will be lowered by ten points."

Student (speaking to a friend): "My history teacher is very strict. He told us that he intends to lower our final grades by ten points if we miss three of his classes."

23. Alex is 35 years old and single; he has had one marriage that ended after two years in divorce. Since his divorce (three years ago), he has dated a number of women but he has been unable to have a lasting relationship with any of them. Here is what Alex says to his counselor:

Alex: "I haven't had any luck with women in the past so how could it ever be any different for me in the future?"

24. Jack, a college student, was taking a college health course. During one class, the instructor addressed the manner in which the AIDS virus (HIV) is spread. The instructor pointed out that there is no evidence that HIV can be spread through casual contact and that it is not likely to be passed through the saliva. However, after class, one of the students, whom Jack had sat next to all semester, told Jack that he had tested "positive" for HIV. Here is what Jack tells himself about the matter:

Jack: "That guy with AIDS was breathing on me all semester. I think that his spit once went on my lips one time when he was talking to me. Now I'll probably get it. I'm dead meat now!"

25. "All you really need to be happy is a lot of money."

26. Here is what Mr. Shackleberry, a high school English teacher, says to a colleague with respect to the fact that more than thirty five percent of his class has, for the past three years, received failing grades:

Mr. Shackleberry: "My job is to teach English, and that's just what I do. If they don't learn the material, then it's their problem, not mine. I'm doing my job and they need to do theirs. All they need to do, if they are to learn English, is to get up off of their rumps and do some reading and studying. That's all there is to it."

27. Todd and Angela have been married for two years. Angela has a 12-year-old daughter, Tammy, from a previous marriage, who lives with the couple. While Angela is very close to Tammy, Todd and Tammy often quarrel. Tammy often tells Todd that he is not her "real father" and that he therefore has "no right" to tell her what to do. Todd, on the other hand, often refers to Tammy as a

"little trouble maker." Recently, Todd and Angela also frequently quarrel with each other. Much of their quarreling centers about Tammy and how, according to Todd, Angela is a "good mother" to Tammy but not a "good wife" to him. Here is one remark that Todd makes to Angela:

Todd: "You haven't been a good wife to me; and Lord knows, I've tried to be a good husband to you."

Angela: "And what's wrong with the way I've treated you?"

Todd: "You're unreasonable, that's what's wrong! I can't even talk to you without your saying something obnoxious or stupid. Let me get right to the point. You just don't treat me right and I'm tired of it. You might be a good mother, but, as a wife, you stink. Does that answer your question or do I have to draw you a picture?"

*28. Here is a conversation between two friends, both of whom have been divorced:

Florence: "Men are all such a bunch of bums."

Vicki: "Why do you say that, Flo?

Florence: "My ex had a roving eye from day one and yours was no different. Women are so different than men. They're not bums like men."

Vicki: "Why's that Flo?"

Florence: "Did you ever cheat on your ex?"

Vicki: "Never!"

Florence: "Did I ever cheat on mine?"

Vicki: "Not that I know of."

Florence: "So isn't that enough proof for you?"

29. "What's the use in living. Nothing ever goes right in my life. Everything that can go wrong always does!"

*30. Julie's 10-year-old son, Mickey has just, today, come down with an intestinal virus. Here is what Julie says:

Julie: "Just yesterday I was saying how well Mickey has been all year--not even one cold. And now today he comes down with this virus. I should have kept my big mouth shut instead of jinxing it."

31. The following occurs between Kim and her husband, Kirk:

Kim: "Honey, why can't we eat out tonight instead of at home? It's been two months since we've gone to a restaurant."

Kirk: "You're always trying to spend our hard earned money. All you ever want to do is eat out. You must think that we're millionaires or something. Can't you let us save a dollar once in a while!?"

32. "All of those southerners are nice to your face; and then when you turn around, they'll stab you in the back."

33. Sam and his wife Kathy are both 36 years old. They have been married for 15 years and have two children, ages 7 and 10. Sam and Kathy both work. However, the couple has almost no social life outside their immediate family. Recently, the two have experienced increasing difficulty in getting along with each other. The children have also began to act out in school and at home. When their marriage counselor suggests that they might try socializing with other couples, here is what Kathy says:

Kathy: "Sam and I have never been able to make friends with other couples very easily. And whatever social life we have had in the past has not lasted very long. I don't know why but it's always been this way and I don't suppose it can ever be any different. I guess we're just destined to be loners for the rest of our lives."

34. John recently lost his job as a salesman and is now faced with having to go out and look for another job. He is quite depressed over his situation. He says the following to his wife:

John: "If only I started my own business when your brother offered to go in partners, I would not be in this situation today."

35. "That stuff that smoking gives people heart and lung diseases is bunk. My grandfather smoked all his life and he lived into his eighties."

36. Here is what one student says about his math teacher, Mr. Morgan:

Student: "I'm flunking math because Mr. Morgan stinks. What a nerd! They ought to fire him. He doesn't know what he's doing."

37. "When I had that car accident, an old Sonny and Cher song was playing on the car radio. Now when that song comes on while I'm driving, I shut it off."

38. Mark and Mia are married and have an infant son. Mark is an executive at a large insurance company. Mia is on maternity leave from her job as a teacher. Both share in the domestic chores, although Mia is presently the primary caretaker of their child. Here is a discussion between the two:

Mia: "Can you get up tonight with the baby?"

Mark: "Can you handle it tonight, honey? I have this important meeting tomorrow and I really would like to get some sleep. It's very important. I can do it tomorrow night instead."

Mia: "You never do anything for me. I ask you to do just this one thing and you refuse to do it. You always have some kind of excuse. Why do I always have to do everything? Can't you ever take anything else seriously besides your work?"

39. Ethel has recently divorced her husband, George, who had been cheating on her. Here is a discussion between Ethel and her close friend, Rachel:

Rachel: "Why don't you let me fix you up with Bernice's brother. I met him once and he seemed very nice."

Ethel: "What does he do for a living?"

Rachel: "Bernice said he's a philosophy professor over at the university."

Ethel: "Forget it. George was a philosophy major too and look what a jerk he turned out to be. I think I'll pass on this one. Maybe you know somebody else."

40. Linda's father had sexually abused her and her two sisters when they were children. Moreover, he had never received any counseling for his problem. Now Linda has a daughter, Jenny, age 7, whom she would like to leave with her father while she and her husband, Ted, vacation in Europe. Here is a discussion between Linda and Ted:

Ted: "I don't really think it's a good idea to leave Jenny with your father. You know what he did to you and your sisters."

Linda: "Don't be ridiculous! Dad would never do anything like that to Jenny. Anyway, that was a long time ago."

41. "I've had a sore throat for three weeks, and it still hasn't gone away. That sounds like throat cancer."

42. Ronald is in the market for a used car. Here is what he says:

Ronald: "I hate looking for used cars. All of those used car salesmen are such slime. Every time one opens his mouth, I expect a lizard tongue to come leaping out at me. Just the thought of going to those guys gives me an anxiety attack."

43. Crystal and Chris have been married for one year. Here is a discussion between them:

Crystal: "You don't love me anymore."

Chris: How can you say that, honey. What's the matter?"

Crystal: "I don't know. You're just different."
Chris: "What's different? Are you talking about my love-making?"
Crystal: "No, but sex isn't everything."
Chris: "Well, then what's wrong? Is it something I said or forgot to do?"
Crystal: "I'm not sure. All I know is that, deep down inside, you really don't love me anymore."
Chris: "Yes I do. I'll always love you, honey."
Crystal: "No, don't deny it. You don't really."

*44. Dina was recently involved in an automobile accident. Another car had failed to stop for a stop sign and struck her car broadside. Fortunately, nobody was seriously injured, including Dina's husband, Sam, who, at the time of the accident, was sitting next to her in the front passenger seat. While the other driver was given a citation, Dina, who was obeying all traffic laws at the time of the accident, did not receive any citation. However, Dina's car was totaled in the accident. In the following, Dina and Sam are about to get into their (other) car to go shopping at the local mall. Here is their conversation:

Dina: "I'll drive, honey."
Sam: "No way Jose'. The last time you did the driving, you had an accident. You better let me do the driving this time."

45. Nick is a teacher who is troubled by the possibility of being laid off due to recent budget cuts in education. Here is what he thinks about the matter:

Nick: "If I'm laid off, I won't be able to get another teaching job. All those years of college will be wasted. My whole life will be a big waste!"

46. Wife: "Are you having an affair with Jenny? I want you to tell me the truth?"

Husband (who just broke it off with Jenny): "No, honey. I swear that there's nothing between Jenny and me."

47. "My ex-wife had a miscarriage. So, my present wife will also have one if she gets pregnant."

48. Here is what one teacher tells another teacher over lunch:

Teacher: "The last time I ate school food--the meat loaf-- I got an intestinal virus the next day. Now I have enough sense to bring my own lunch!"

FURTHER PRACTICAL EXERCISES FOR CHAPTER TWO

1. Dividing in smaller groups, let each group create and take turns *role playing* (before the whole class) at least one practice scenario containing at least one **Fallacy of Fact Gathering.** The class may then critique the scenario, identifying the specific fallacy or fallacies committed and discussing how the characters in the scenario could have more effectively dealt with their situation. (*A Further Option*: After the class finishes its critique, the group may then "replay" the scenario, however this time omitting the previous fallacy or fallacies, dealing more effectively with the original situation.)

2. Collect at least 5 examples of at least five *different* **Fallacies of Fact Gathering**. You may find these in newspaper editorials, magazine articles, advertisements, TV or radio commercials, political addresses, books, essays, etc. You may also use personal experiences you have had such as what you or somebody else said under a given set of circumstances. For each fallacy you find, write down the name of the fallacy and briefly describe why you believe it to be that fallacy. If you have *BELIEF-SCAN* you may use the

program to support or establish your findings. Include in your project package copies of any supporting newspaper clips, advertisements, etc. that you have used (where available). Also attach any supporting *BELIEF-SCAN* re- ports. Be prepared to present and discuss findings in class.

3. Select an essay or editorial (from any source you want) written on a controversial subject, and analyze it for **Fallacies of Fact Gathering**. Write a brief evaluation of it's strengths and weaknesses in the light of your findings. If you have *BELIEF-SCAN,* you may use it to check for these fallacies.

CHAPTER THREE
LOGICAL FALLACIES

WHAT ARE LOGICAL FALLACIES?

Logical Fallacies involve mistakes in *structure or shape* of your reasoning. For example, they involve mistakes in using words such as "either. . .or. . ." "if. . .then. . ." and "not." These fallacies are discussed in what follows.

1. THE FALLACY OF THE WHOLE

This fallacy occurs when you assume that because a part of something is a certain way then the *whole* thing must also be that way. Although wholes may sometimes be like their parts, it is a mistake to think that they must be so. For example, Ali makes this mistake when she believes that "If my husband and I are not having good sex, then our marriage is not good." Although a sexual relationship may be an important part of a marriage, it is not the only one; and it is possible that the marriage may otherwise be quite good.

The Fallacy of The Whole is sometimes confused with Hasty Generalizing, a fact gathering fallacy discussed in the previous chapter. However, in the case of Hasty Generalizing you generalize about *each and every* member of a group on the basis of what is true about *some* of its members. On the other hand, The Fallacy of The Whole does not involve making any claim about each and every member of a group. Rather, it involves making a claim about *the whole itself.* For example, The Fallacy of the Whole is committed when you conclude that the cake must be tasty because you like one of its ingredients, namely brown sugar. In drawing your conclusion you are *not* claiming that, because you like brown sugar, you like each and every other ingredient. This would be a Hasty Generalization. Instead you are concluding that *the cake itself*, as distinct from any of its ingredients, must be good.

2. THE FALLACY OF THE PART

The Fallacy of the Part is just the opposite of the Fallacy of the Whole (which is discussed above). It is to suppose that what's true of the whole must also be true of the part. For example, Henry makes this mistake when he believes that "If my wife and I have a good marriage, then we must also have good sex." Just as you can't say what the whole marriage is like from just a part of it, you can't say what a part of a marriage is like from the whole.

In order to keep the Fallacies of The Whole and Part straight, it may be helpful to look a what your *conclusion* is about. More specifically, when you commit The Fallacy of The Whole your conclusion will be about *the whole*. On the other hand, when you commit The Fallacy of the Part, your conclusion will be about *the part*. For example, if you conclude that a certain college is itself outstanding because it has an outstanding physics department, you are committing The Fallacy of The Whole. This is because you are *concluding* something about *the whole college*. On the other hand, if you conclude that the physics department of a certain college must be outstanding because the college itself is outstanding, then you are committing The Fallacy of The Part. This is because you are *concluding* something about *a part* of the college, namely its physics department.

3. ARGUING IN A CIRCLE

Arguing in a Circle occurs when you try to defend your belief with the same belief you are trying to defend. For example, John argues in a circle in the following discussion between him and his counselor:

John: "My wife acts like a tramp."
Counselor: "Why do you think so?"
John: "She flirts with other men."
Counselor: "Why do you say that?"
John: "You know, she comes on to them."
Counselor: "But why do you think she comes on to other men?"
John: "I told you; she acts like a tramp; you know what I mean."

In the above discussion, John finally defends his belief that his wife acts like a tramp by repeating that she acts like a tramp!

4. THE FALSE "EITHER OR"

The fallacy of The False "Either Or" occurs when you treat two choices as the only ones when in fact there are other possible choices. For example, Tammy makes this mistake when she believes that "Either George loves me or he hates me." Between love and hate are many other emotions. For instance, George might just *like* Tammy without loving or hating her. Similarly, Ned makes this mistake when he believes "If my girlfriend doesn't have sex with me, then she doesn't love me." Ned's girlfriend may simply not believe in premarital sex!

Keep in mind that reality is often not "black or white." In your attempt to "pigeonhole" reality into just two categories, you are likely to miss important distinctions and alternatives; and this could, in turn, have self-defeating practical results. For example, in seeing things in terms of "love or hate" exclusively, Tammy may miss an opportunity to cultivate a close friendship, as distinct from a romantic relationship, with George. And Ned may break up with his girlfriend on the false grounds that she really doesn't love him.

5. GRAY AREA FALLACY

The Gray Area Fallacy occurs when you suppose that because there are "gray areas" there are no clear cases. For example, Elizabeth believes that, "If you can't tell just where showing affection for your child ends and child sexual abuse begins, then it's not really right to charge my husband with sexually abusing my daughter." Although there may be some gray areas where it is not clear whether or not something is child sexual abuse, that does not mean that there are no clear cases of it for which a person can and should be held responsible. Elizabeth's husband's conduct may have crossed over the line between simple fatherly affection and sexual abuse. For example, such a line is crossed when the father fondles his adolescent daughter's breasts.

Keep in mind that, while reality is often not "black or white," still neither is it always "gray." Like the False "Either Or," seeing things exclusively in terms of the gray areas can also have self-defeating practical results. For example, in focusing on the gray areas, Elizabeth may allow her husband to get away with abusing their daughter.

6. INCONSISTENCY

Inconsistency occurs when you believe two things at once that cannot both be true at once. For example, "I love my father" and "I don't love my father"; "I can't survive without my husband" and "I can do fine on my own without my husband."

Look, for example, at the following discussion between Barbara and her counselor:

Barbara: "I hate my mom."

Counselor: "Why?"

Barbara: "A few years ago, she divorced my dad and moved away from us. I hardly get to see her now."

Counselor: "Why do you hate your mom because of that?"

Barbara: "It still hurts me to think she would leave like that."

Counselor: "Why does this still hurt you?"

Barbara: "I just wish she would come home so that things could be the way they were."

Notice that Barbara says both that she hates her mom and that she wishes she would come home. Barbara seems to be longing to get close to somebody she says she hates. There is a conflict here that Barbara needs to work through. For example, maybe Barbara does not really hate her mother but has not yet come to terms with what her mother has done. The seeming inconsistency between Barbara's claims is a cue that she has not yet gotten clear about her own feelings and may need to explore things further.

Inconsistencies sometimes take the form of *double standards*. These arise when you treat *relevantly* similar cases differently. For example, Ryan, an employer who prefers male employees to female ones, gives Pete a raise but does not also give Sue a raise. Given no *relevant*, work-related distinctions between Pete and Sue (years of service, performance, etc.) upon which to base the differential treatment, Ryan is guilty of a double standard. It is this type of inconsistency that defines the concept of *unfair or unjust discrimination*.

Double standard inconsistency is often submerged under a Smoke Screen of emotional language with no real factual basis to warrant the distinction. For example, according to Andy, "I may be frugal when it comes to spending money, but James is nothing but a penny-pinching cheapskate." And Harry says, "I admit I have a hearty appetite, but Harvey makes a pig of himself." When the smoke settles, it becomes clear that Andy and James have similar money- spending habits; and Harry and Harvey help themselves to similar portions.

7. THE VICE VERSA FALLACY

Look, for example, at the following two statements:

(1) *If my wife and I have a good marriage then my wife and I do not cheat on each other.*

(2) *If my wife and I do not cheat on each other then my wife and I have a good marriage.*

Notice that statement (2) reverses the order of statement (1). If you think that statements (1) and (2) mean the same thing, then you are committing the Vice Versa Fallacy. In fact, each statement means something different. Statement (1) says that, "If my wife and I have a good marriage, this means that my wife and I don't cheat on each other." Statement (2) however says that, "If my wife and I don't cheat on each other then this means that my wife and I have a good marriage." Having a good marriage might mean not cheating, but not vice versa. Just because you don't cheat doesn't mean you have a good marriage. There are some couples who remain loyal to one another even though they are very unhappy together. These couples do not have good marriages. You can avoid the Vice Versa Fallacy by paying close attention to the order in which you are saying something. Failure to do so may lead you to say things you really don't accept or to believe things you really have no good reason to believe.

8. FALLACY OF DERIVING "IF" FROM "THEN"

This fallacy arises when you conclude that because the THEN-clause of an "If...Then..." statement is true, then the IF-clause **must** also be true. For example, suppose you know that the following "If...Then..." statement is true:

If Sam is sterile then he has not impregnated his wife.

Well, suppose that you also know that Sam *has not* in fact gotten his wife pregnant so that the THEN-clause of the above statement is true. Is this information sufficient for you to conclude that the **IF**-clause must also be true, namely that Sam must be sterile? Clearly not, because there are a number of other factors that could also explain the lack of pregnancy (including the possibility that the couple has not even tried to get pregnant!). Thus, if you were to conclude that Sam *must* be sterile just from the given information, then you would be committing the fallacy of Deriving "If" from "Then".

The "must" here is important. If you simply concluded from the given information that Sam *might* be sterile, you would not be committing the fallacy in question. The fallacy would arise only if you thought that it *necessarily followed* that Sam was sterile. In general, it arises whenever you think the "If" necessarily follows from the "Then."

You can avoid the fallacy of Deriving "If" from "Then" if you keep the following in mind. Whenever you believe an "If...Then..." statement, that is, one of the form

IF **This** THEN **That**

do *not* conclude from it that the "If" part (the **This**) must be true just because you also happen to think that the "Then" part (the **That**) is true.

However, while the "If" does not necessarily follow from the "Then," the "Then" *does* necessarily follow from the "If". For example, given the truth of the following two statements,

If John is married then he will not ask Jenny out on a date

and

John is married,

you can be sure that John will not ask Jenny out on a date *but only on the assumption that the above two statements really are true.* Since John might sometimes cheat on his wife, the first statement might not be true; in which case, the conclusion that John will not ask Jenny out on a date would also be in question.

The fallacy of Deriving "If" from "Then" resembles the Vice Versa fallacy but there is one major difference. When you commit the Vice Versa Fallacy, you simply derive "If **That** then **This**" from "If **This** then **That**," thereby switching the order of the "If" and "Then." In the case of Deriving "If" from "Then" your conclusion turns out to be the "If" by itself, *without* the "Then." For example, in the previous example, the conclusion was "Sam is sterile." In the case of the Vice Versa fallacy, the conclusion would have been "If Sam has not impregnated his wife, then he is sterile."

FALLACY OF DERIVING "THEN NOT" FROM "IF NOT"

This fallacy arises when you conclude that because the IF-clause of an "If...then..." statement is false, the THEN-clause *must* also be false For example, suppose that you are driving and approaching an intersection that has a traffic light. You notice that the light is not red (it is in fact green). Since the light is not red, you feel sure that you do not have to stop. So you enter the intersection. Sound familiar?

Let's look at your reasoning in this example: since you know what a red traffic light signifies, you know that the following "If...then..." statement is true:

If the traffic light is red then I have to stop.

You also know that the traffic light is *not* red so that the IF-clause of the above statement is false. Because the IF-clause is false, you conclude that the THEN-clause must also be false, namely that you do *not* have to stop. This would be the fallacy of Deriving "Then Not" From "If Not".

This is a fallacy because there may be other reasons for you to stop even if you do not have a red light. For example, suppose that a young child runs into the road to get a ball. You do not have a red light so you enter the intersection, striking and killing the child. In such a case the conclusion that you do not have to stop (the "Then Not") does not *necessarily* follow from the light's not being red (the "If Not").

The fact that the light is not red, does however support the conclusion that you *might not* have to stop. But this is also consistent with the possibility that you *do* have to stop, for example, for a pedestrian, a traffic jam, a malfunctioning light, or even an amber about to turn red. This is why it is a good idea to slow when approaching an intersection even when you do not have a red light. By so doing you will give yourself time to see if the coast is clear. Those who try to "make" or "beat" a light commit the fallacy in question by failing to see that a non-red light (amber or green) does not *necessarily* mean that they do not have to stop.

You can avoid the fallacy of Deriving "Then Not" from "If Not" if you keep the following in mind. Whenever you believe an "If...Then..." statement, that is, one of the form

IF **This** THEN **That**

do *not* conclude from it that the "Then" part (the **That**) must be false just because you also happen to think that the "If" part (the **This**) is false.

However, while the "Then Not" does not necessarily follow from the "If Not," the "If Not" *does* necessarily follow from the "Then Not". For example, suppose that the following statement is true:

If a person shares IV drug needles then he or she is at high risk of getting AIDS

Since Lenny is *not* at high risk of getting AIDS, it *does* necessarily follow that he does *not* share IV drug needles ("If Not"). On the other hand, even if Don does *not* share IV drug needles, it does *not* necessarily follow that Don is *not* at high risk of getting AIDS. This is because Don may have other factors that make him high risk, for example, that he is promiscuous.

REVIEW DEFINITIONS: LOGICAL FALLACIES

1. **THE FALLACY OF THE WHOLE** =*Df.* Assuming that, because a part of something has a particular characteristic, the whole thing must also have that same characteristic.

2. **THE FALLACY OF THE PART** =*Df.* Assuming that, because a whole has a particular characteristic, a part of it must also have that same characteristic.

3. **ARGUING IN A CIRCLE** =*Df.* Trying to justify or defend a belief by appealing to the very same belief that you were trying to justify or defend in the first place.

4. **THE FALSE "EITHER OR"** =*Df.* Presenting a situation in such a way that you see only two alternatives while overlooking others.

5. **GRAY AREA FALLACY** =*Df.* Assuming that, because there are unclear cases of something, there cannot, therefore, be any clear cases of it.

6. **INCONSISTENCY** =*Df.* Asserting (saying, thinking) two things that cannot possibly both be true at the same time.

7. **THE VICE VERSA FALLACY** =*Df.* Reversing the order of an "If *this* then *that*" statement and taking it to mean the same as

"If *that* then *this*".

8. **FALLACY OF DERIVING "IF" FROM "THEN"** = *Df.* Given an "If *this* then *that*" statement, thinking *this* **must** be true because you think *that* is true.

9. **FALLACY OF DERIVING "THEN NOT" FROM "IF NOT"** =*Df.* Given an "If *this* then *that*" statement, thinking *that* **must** be is false because you think *this* is false.

PRACTICE SCENARIOS FOR CHAPTER THREE

EACH OF THE CASES PRESENTED BELOW CONTAIN AT LEAST ONE OF THE FALLACIES DISCUSSED IN THIS CHAPTER. FOR EACH CASE, NAME AT LEAST ONE FALLACY IT CONTAINS AND EXPLAIN HOW THE PERSON WHO COMMITTED THE FALLACY MIGHT HAVE AVOIDED IT. THAT IS, EXPLAIN HOW THIS PERSON COULD HAVE MORE EFFECTIVELY DEALT WITH HIS OR HER SITUATION.

1. John has been married to Fran for twelve years. Recently, Fran has announced her intention to file for a divorce. Here is what John tells his marriage counselor:

John: "Fran has been a very important part of my life. If my marriage ends, then my whole life might as well have ended too."

2. Gloria and her husband are having problems relating to one another. Here is a discussion from a session between Gloria and her counselor:

Gloria: "My husband promised to take care of the kids on Saturday and he didn't. Instead he worked the entire day. I think people should either keep the promises they make or just not make them in the first place. Wouldn't you agree?"

Counselor: "Did you ever break a promise?"

Gloria: "Well, yes. My kids had asked me to take them shopping for toys. I promised them that I would take them this evening. But, as you can see, I' m sitting here with you instead. It was really important for me to see you, though. I just did what I thought was right, you know?"

*3. Elfidge and Abigail have lived together for the past three months in order to see if they are both ready to make a commitment to marriage. In that time, the two have experienced difficulties in getting along with one another. Abigail has accused Elfidge of not being a very good conversationalist. On the other hand, Elfidge has maintained that Abegail complains too much about little things in her life. Elfidge has also told Abigail that if he is to be happy living with her, then she has to stop her frequent complaining and that he will try to be a better conversationalist. In the past two weeks Abigail has, indeed, stopped her complaining. However, she also has stopped speaking openly to him. Now, when Elfidge addresses her, she either nods or gives him single word answers. Elfidge has just complained to her about her new behavior. Here is Abigail's response:

Abigail: "You told me that if you are to be happy living with me, then I have to stop complaining so much. Well I haven't complained one bit in the past two weeks and now you're still not happy. I guess you don't know yourself how you want me to act toward you."

4. Samantha has recently found her husband in bed with her best friend. Here is a discussion between Samantha and her counselor:

Counselor: "How do you feel about your husband having had an affair with your best friend?"

Samantha: "I really don't care one way or the other anymore. At first I admit I felt like killing them both but now it just doesn't matter to me. I feel nothing at all now. Anyway, maybe he'll catch some disease from that two-timing whore. I would have the last laugh then."

5. Here is what one young women anxiously awaiting a blind date says to her girlfriend:

Jenny: "My date for tonight is on the high school football team. That team has lost every game this season; so he must be a loser too."

6. During a class lecture, a student raises her hand and asks the teacher if she can go to the bathroom:

Student: "Mr. Mann, can I go to the bathroom now?"
Mr. Mann: "No."
Student: "But I really need to go badly. Why can't I?"
Mr. Mann: "Because I said no. Now sit down."

7. "My wife is not going to push me around like that. If anyone is going to do the pushing, it will be me, not her."

*8. Freda has been married to Bob for 19 years. She is now 40 years old and she has become extremely self-conscious about her age and about getting older. Recently, Bob's boss met Freda for the first time at a company party, commented on how attractive she was, and asked Bob her age. Bob, in turn, told his boss that Freda was 40. Here is a discussion that takes place between Bob and Freda after their return home from the party:

Freda: "You really shouldn't have told your boss how old I was when he asked you."
Bob: "Why not? What should I have done?"
Freda: "You should have lied to him or something."
Bob: "But why should I have lied to him?"
Freda: "Because my age is my business and nobody else's. It's just not right to go telling other people my business."

Bob: "But why is it just your business? Why can't anyone else know how old you are? Are you ashamed of your age?"

Freda: "Of course not!"

Bob: "Then why can't anyone else know your age?"

Freda: "I told you already, didn't I?" You just shouldn't have told him. That's all. Now leave me alone!"

9. Hal has recently lost his job. Here is what he tells his counselor:

Hal: "Either a man is successful or he's a failure. I can't call myself successful anymore. So you know just where I stand."

10. "It's hard to tell the difference between deliberately doing unhealthy things and attempted suicide. So, as far as I'm concerned, people who smoke cigarettes or live on junk food are just trying to commit suicide."

11. Pete's friend, Jim, has recently complained to Pete that he is "short on cash" and that he does not have enough money "to make ends meet." Pete, however, does not believe Jim. Here is what Pete tells his girlfriend:

Pete: "If a single guy is to live comfortably in this day and age then he needs to make at least thirty grand a year. Jim is single with no family to support or alimony payments to make, and he makes that kind of money easy; so he must live comfortably. He's got what to complain about!"

12. Rodney has been indicted for selling drugs. He has asked his girlfriend, Becky, to take the stand and to lie to the court under oath for him. Becky, however, has refused to do this for Rodney. Here is what Rodney tells Becky after she refuses his request:

Rodney: "Either you're willing to do anything you can for me or you really don't care about me. So, why don't you just go away and leave me alone. Go on, get out of my life for good!"

*13. Here is a discussion between two close friends, John and George:

John: "I believe that a man has got to live by the Bible."

George: "You mean a man shouldn't do anything that the Bible forbids?"

John: "That's right. Like those homosexuals. They're all sinners against the Lord."

George: "What about adultery, John? Didn't you cheat on Claire a couple of times? I remember two times, once with Alma and once with Hannah, and Claire never even found out. Isn't that right, John? Doesn't that make you a sinner too?"

John: Hell no! That's a different story. Sometimes a man has got to get it out of his system. Claire and I have a fine marriage now and I owe it all to those little flings with Hannah and Alma."

14. "If he's so smart, how come he didn't know the answer to that question?"

15. Hubert was a student in Mr. Carlswood's biology class. His final test average was fifty-five. He makes this comment to another student:

Hubert: "That Carlswood told the class that in order to pass the course, we had to have good attendance. Well, I came to almost every class and he still flunked me. I don't think a teacher should mislead his students like that, do you?"

16. Father helping his son with his school science project:

"If you're not going to do it my way, then don't even bother doing it at all."

17. "My car keeps stalling out; and that's just what would happen if the fuel pump was bad. So it must be a bad fuel pump."

*18. Dave is looking for a family physician. Here is a discussion he has with his friend, Sid:

Dave: "What do you know about Dr. Loomis?"

Sid: "I don't know much about him. I do know that he is on staff at Morris Memorial."

Dave: "Really? That hospital stinks. Loomis must also stink if he's a member of that hospital. Forget him. Know anyone else who's good?"

19. Pat and Pauline have been married for 17 years. They are both employed and have two children, ages 11 and 6. Ever since the birth of their second child, the lives of the two have revolved, for the most part, around work and what they take to be the needs of the children. Both Pat and Pauline admit that they are unhappy and that there is "something missing" from their marriage. The two have been seeing a marriage and family therapist who has recommended that the two get out more together without the kids. In one session, here is what Pat tells the therapist:

Pat: "You said that if our marriage is going to improve, then we need to spend more time together, alone, away from the kids. Well, in the past two months we've done what you said and still our marriage has not really gotten any better. When we're out together it seems alright but when we get home the honeymoon's over. So that's not our problem. You were just wrong. Spending more time together without the kids doesn't mean our marriage will improve."

20. Here is what one student says to his teacher, Mr. Seymour, who is chewing gum above a sign that reads "No Eating, Drinking or Gum Chewing in This Class."

Student: "Mr. Seymour, how come you chew gum and we can't?"

Mr. Seymour: "I'm the teacher. That's why!"

21. Missy and Rupert are currently living together out of wedlock. Rupert wants to get married and has recently proposed to Missy. However, Missy is contented with the present living arrangement and is not convinced that marriage is such a good idea. Here is a discussion between Missy and her friend Sandra:

Missy: "Rupert told me that if I marry him he will love and cherish me until death do us part."

Sandra: "Oh how romantic."

Missy: "Really? I don't think so. Suppose we never do get married and instead just live together. Then that must mean he *won't* love and cherish me until death do us part. What kind of love is that anyway! He makes love sound like a legal contract!"

22. Sam is a chain smoker. Here is a discussion between him and his counselor:

Sam: "I need a cigarette to relax me when I get uptight."

Counselor: "Why do you think you need a cigarette to relax you in those situations where you feel uptight?"

Sam: "I have to have one in order to relax. I can't relax unless I have one."

Counselor: "Why do you think you can't relax without one?"

Sam: As I said before, I need one or else I can't relax!"

*23. Joe has developed a drinking problem. His wife, Anne, suggests to him that he might be an alcoholic. Joe, however, insists that he is only a "social drinker." Here is one of their conversations:

Joe: "Can you tell where social drinking leaves off and alcoholism begins? How much do you have to drink before you're no longer a social drinker? Can you even tell me that?"

Anne: "I admit that sometimes it's hard to tell the difference between them."

Joe: "Then how can you say that I'm an alcoholic. You might as well say that I'm just a social drinker. That's what I really am, so why don't you just admit it."

24. This morning Maria's son, Teddy, complains of a headache and dizziness and asks if he can stay home from school. Here is what Teddy's mother says to him:

Maria: "If you had a fever then you would be sick. But I took your temperature twice and you don't have fever. You're 98.6 right on the dot. So, you're not really sick.
Get your tail out of bed and get ready for school you little con artist."

25. Here is what one teacher tells a student who asks for a unit review:

Teacher: "You should know this material. Either you know it by now or you never will!"

26. "The world is a bad place because bad things happen in it."

27. Parent: "I told you to get to bed right now."

Child: "But why do I have to go to bed right now? Can't I stay up a little while longer?"

Parent: "No."

Child: "But why can't I?"

Parent: "Because I told you to get to bed right now, that's why."

28. Jane and Howard are boyfriend and girlfriend. Here is what the say to each other after having had a quarrel:

Jane: "If you apologize to me then I'll apologize to you."

Howard: "Alright, go ahead and apologize to me and then I'll apologize to you."

29. "There's no way to tell exactly how overweight a person has to be in order to be called `morbidly obese', so how can you stand there and tell me that that's what I am."

30. Here is what one 9-year-old boy tells his friend:

"My mom said she would buy me a Sega CD if my dad said it was OK. But I know he's gonna say no; he's always sayin how video games are for airheads; so I'm sure my mom won't buy one for me either."

FURTHER PRACTICAL EXERCISES FOR CHAPTER 3

1. Dividing in smaller groups, let each group create and take turns *role playing* (before the whole class) at least one practice scenario containing at least one **Logical Fallacy**. The class may then critique the scenario, identifying the specific fallacy or fallacies committed and discussing how the characters in the scenario could have more effectively dealt with their situation. (*A Further Option*: After the class finishes its critique, the group may then "replay" the scenario, however this time omitting the previous fallacy or fallacies, dealing more effectively with the original situation.)

2. Collect at least 4 examples of at least 4 *different* **Logical Fallacies**. You may find these in newspaper editorials, magazine articles, advertisements, TV or radio commercials, political addresses, books, essays, etc. You may also use personal experiences you have had such as what you or somebody else said under a given set of circumstances. For each fallacy you find, write down the name of the fallacy and briefly describe why you believe

it to be that fallacy. If you have *BELIEF-SCAN* you may use the program to support or establish your findings. Include in your project package copies of any supporting newspaper clips, advertisements, etc. that you have used (where available). Also attach any supporting *BELIEF-SCAN* reports. Be prepared to present and discuss your findings in class.

3. Select an essay or editorial (from any source you want) written on a controversial subject, and analyze it for **Logical Fallacies**. Write a brief evaluation of it's strengths and weaknesses in the light of your findings. If you have *BELIEF-SCAN,* you may use it to check for these fallacies.

CHAPTER FOUR
FALLACIES OF MEANING

WHAT ARE FALLACIES OF MEANING?

Fallacies of Meaning are committed when words are used in *misleading or unclear* ways. The following is a discus- sion of each of these fallacies.

1. PERSONIFYING

The fallacy of Personifying occurs when you speak about things that are not really persons as if they actually were persons. For example, "Let nature take care of things" gives "nature" the human ability to care for things; the expression "Life has been good to you" gives "life" the ability to do good; the statement "I have always been a victim of circumstances" gives "circumstances" the ability to victimize. Notice that mistakes of this sort may lead you not to *take responsibility* for your own life, relying instead on the goodness of life itself or on the power of circum- stances to victimize.

Common objects may also be personified and used to sidestep responsibility. For example, when Jack's two cars brake down at the same time, he curses them for conspiring against him! But when was the last time Jack changed the oil or gave his cars a tune-up!? It may be "easier" for Jack to blame the car than to accept responsibility for maintaining them. But what may be easier to do in the short run may in the long run be self-defeating.

There are numerous and often subtle ways in which we humans may, without realizing it, engage in Personifying as a way of sidestepping responsibility. A major practical problem with doing so is that you thereby lose the opportunity to *rationally* determine or control the course of events in your life. Thus, for example, by relying on "fate" or "destiny" to guide you through life, you fail to exercise your own rational judgment. Since there are typically more ways in which events can randomly "go wrong" for you (that is, more ways in which you would not want them to go), by trusting to "fate" or "destiny" it is more likely than not that you will, in the end, be disappointed.

2. SWITCHING MEANINGS

The fallacy of Switching Meanings occurs when you confuse two or more different meanings of the same term. For example, George says "If my wife doesn't make love to me, then she must not love me." In George's statement, the word "love" occurs twice. However, the first occurrence of "love" refers to sex whereas the second occurrence refers to an emotion of great caring for a person. While having sex can be a way of expressing love in the second sense, it is not really necessary for such love. For instance, even those who are unable to have sex due to illness may be in love in the second sense. George therefore has "switched" the meaning of the word "love" and, so, falls into the fallacy of Switching Meanings.

3. VAGUE TERMS

Vague terms are words not having clear meanings or which need to be defined or clarified. For example, Jessica says, "I wish I were married to a normal man." In this case, Jessica's dissatisfaction with her husband cannot be dealt with unless she is clear about what she means by the word "normal." For example, when asked what she means, Jessica says "A normal man is one who has sex with his wife more than once a month." With this understanding, the couple's sex life can now be discussed.

In addition to single words, *phrases* may also need to be clarified. For example, Sue says "My mom lives in her own little world." What Sue means by "her own little world" may not be clear. For example, in response to her counselor's request for clarity, Sue adds, "My mom never sees anything from anybody else's point of view except her own." Now Sue's attitude toward her mother can be discussed.

Since terms can sometimes have more than one meaning, it is not always clear which of these different meanings is intended by the speaker. For example, Karl says "I have been on my own since ten years old when my father died." Without knowing much about Karl's past, this statement could mean at least two different things: it could mean "I have lived with little or no parental or adult supervision since the age of ten when my father died"; but it could also mean "I have not been supported *emotionally* by anybody since the age of ten when my father died." (Maybe Karl means both of these things!) In any case, Karl's statement is vague and its intended meaning needs to be clarified.

4. OPPOSITE MEANING

The fallacy of Opposite Meaning occurs when you verbally express exactly the opposite of what you really mean. For example, Ben is ill and is unable to attend a party to which he and his wife, Claire, have been invited. What Ben really wants is that Claire stay home with him and keep him company; instead, however, he tells Claire "Why don't you go on without me; I'll be alright." "Are you sure, honey?" says Claire. "Yeah, its OK, go ahead" responds Ben. If Claire does not pick up on the opposite message Ben is sending through nonverbal means such as body language and voice intonations, she may go on to the party without him. Upon returning, Ben may express disapproval at Claire's having gone to the party without him, as if she should have known what Ben *really* wanted! Of course, Ben could have communicated his wishes more effectively to Claire, in the first place, had he simply *said what he meant* rather than just its opposite.

5. SELF-EVIDENCE FALLACY

Something is "self-evident" when you can see that it *must,* necessarily, be true just by *thinking* about it for a moment. No other evidence is needed. All you have to do is to understand the *meanings* of the words that are used. For example, a moment's thought will tell you that "All triangles have three sides" and that "All red things are colored." If you know what the words in these statements mean then you will see that these statements must be true.

The Self-Evidence Fallacy occurs when you think, or act as if, something is self-evident, and therefore *must* be true, when it really is not. For example, Harriet commits this fallacy when she insists that she must always let her husband have the final say in all family decisions because "the man must make all the decisions just because he is the man." While a triangle must have three sides in order to be a triangle (that is part of what it *means* to be a triangle), a man does not have to make all the decisions in a household in order to be a man (that is *not* part of what it means to be a man). While Harriet may choose to let her husband be the chief decision maker, she does not have to do so because he is a man. By committing the self-evidence fallacy, Harriet has lost a sense of her own freedom and responsibility for choosing to remain subordinate to her husband.

In order to avoid the Self-Evidence Fallacy, you should pay careful attention to the meanings of the words you use. For example, Joan thinks that her 4-year-old son is a liar because he lied to her about having taken a dollar out of her pocketbook. Joan here commits the Self-Evidence Fallacy because she thinks that telling a lie necessarily makes a person a liar. This, however, is not self-evident. In fact, it is not even true. If it were then all (or almost all) of us would be liars since who amongst us has not ever told even a single lie? By definition, a liar is somebody who makes a *habit* out of lying. Since Joan's son is not necessarily in a habit of telling lies, he is not necessarily a liar, even if he has told a lie. Joan could have avoided the Self-Evidence Fallacy had she paid attention to the meaning of the words she used, in particular, the word "liar." She might have also, thereby, dealt with her son more appropriately. For example, she called him a "liar" rather than simply explaining to him the importance of speaking truthfully and of staying out of mommy's pocketbook.

In general, you can guard against committing this fallacy by asking yourself such questions as "Is that *self-evident*?" "Is that *necessarily* so?" "Do I really *have to* do this?" "Could that *possibly* be false?" "Must this *always* (without any ifs ands or buts) be true?" "Is that what it *means* to be such?" Of course, asking these questions may not be enough. Before saying yes or no to them, you should also give them careful consideration.

REVIEW DEFINITIONS: FALLACIES OF MEANING

1. **PERSONIFYING** =*Df.* Thinking or speaking of non-human objects as though they had human characteristics or powers.

2. **SWITCHING MEANINGS** =*Df.* Using a word to mean one thing but then misleadingly going on to use the same word again to mean something quite different.

3. **VAGUE TERMS** =*Df.* Trying to communicate with another person by using words or phrases that lack a clear meaning.

4. **OPPOSITE MEANING** =*Df.* Saying something that is not consistent with what is really meant.

5. **SELF-EVIDENCE FALLACY** =*Df.* Mistakenly assuming that a certain statement has to be true and that I do not need any proof for it beyond simply understanding what the statement is saying.

PRACTICE SCENARIOS FOR CHAPTER FOUR

EACH OF THE CASES PRESENTED BELOW CONTAIN AT LEAST ONE OF THE FALLACIES DISCUSSED IN THIS CHAPTER. FOR EACH CASE, NAME AT LEAST ONE FALLACY IT CONTAINS AND EXPLAIN HOW THE PERSON WHO COMMITTED THE FALLACY MIGHT HAVE AVOIDED IT. THAT IS, EXPLAIN HOW THIS PERSON COULD HAVE MORE EFFECTIVELY DEALT WITH HIS OR HER SITUATION.

1. Victor and Victoria have been married for one year and recently have separated. Both Victor and Victoria miss one another and each secretly hopes the other will make the first move so that they can get back together. One day Victor pays Victoria a visit. Victor asks Victoria to get back together and here is her response:

Victoria: "How dare you think that I will come running back to you the minute you snap your fingers! Please leave this instant."

2. John's wife has recently divorced him and he is de- pressed. He says the following to his counselor:

John: "Ever since I can remember, life has played cruel tricks on me. I thought I had a wife who really cared, but that was just another one of life's teases. I was just set up just to be knocked down."

*3. James and his girlfriend, Sherry, had a quarrel. When Sherry told James that she never wanted to see him again, he began to cry. In what follows, James discusses the matter with his good friend, Curt:

James: "I just don't know what came over me. When she said we were through, I just couldn't hold back the tears. I cried just like a woman. A man is not supposed to cry like that. What kind of a man am I anyway. Imagine what Sherry thinks of me now!"

Curt: "Maybe Sherry will change her mind. Some women like sensitive men."
James: "Sensitive men, my eye. A man who bawls like that is no man at all."

4. Here is how Karen describes her husband to her counselor:

Karen: "My husband has what I call his "dark moods." When he has them, I try my best to keep my distance from him."

5. Bob: "Are you angry at me?"
Bob's Girlfriend (speaking with sarcasm): "Me? Angry? Now why in the world would I be angry at you."

6. Lyle has a heart condition. His wife Laura thinks that Lyle takes advantage of her by asking her to do too many things for him that he can do by himself. Here is a conversation between the two:

Lyle: "Where's my supper?"
Laura: "It's in the oven. Why don't you take it yourself."
Lyle: "Can't you serve it? I need to sit down and rest. I feel sick."
Laura: "Sick again. You sure are sick! Sick, Sick, Sick! If you're so sick, why don't you see a psychiatrist."

7. "A real wife must have supper on the table before her husband comes home."

*8. Carla has chronic bronchitis, which is aggravated by inhaling cigarette smoke. Recently, she has accepted a new job as an Executive Secretary to The Vice President of a large corporation. However, she later finds out that The Vice President smokes. Here is a discussion between the two on the first day of Carla's employment when she first finds out about her new boss's smoking habit:

Vice President: Do you mind if I smoke while I dictate this letter?

Carla: "You smoke."

Vice President: "That's right. Several packs a day. Been smoking since I'm a kid. Can't seem to kick the habit. Do you mind?"

Carla: "Well [she then coughs], yes. I mean no, go ahead. I don't mind."

Vice President: "Are you sure?"

Carla: "Yes [she coughs again], it's OK."

Vice President: "Good because my last secretary wouldn't let me smoke around her. Let's get started, shall we?"

Carla: "Yes, let's."

9. Laura: "Johnny is always there when I need him. He is really a beautiful person."

Laura's Friend: "Beautiful? He might be a nice guy and all, but he's not much to look at."

10. Here is a discussion between two students, Justin and Larry, who meet in passing:

Justin: "How's your semester been treating you?"

Larry: "Not very well. That algebra is really wicked. It has been playing havoc with me from day one."

*11. In what follows, Shelly comments to her friend, Stacey, about Paul, a man she dated last Saturday night for the first time:

Stacey: "What did you think of Paul?"

Shelly: "He seems very nice but he has very strange ideas about life."

Stacey: "Strange?"

Shelly: "You know, weird, peculiar, way out. I really don't think I could ever go with a man like that. I think I would be better off with someone who is more down to earth, someone more like the way I am. You know?"

Stacey: "Then why don't you call Bob. He's a real down to earth guy. If you ask me, he's a real catch."

Shelly: "I just might do that. What's his number?"

12. "It took me so long to find the ketchup in my refrigerator. I finally found it hiding behind the bread."

13. Ken: "Let's do something fun."

Ken's Mate: "Sure, what would you like to do?"

Ken: "I don't know, let's just have fun."

14. Clark is a college student who is considering dropping out of college in order to support himself. While he has some wealthy relatives, Clark does not want to ask them for any financial help. Here is a conversation he has with one of his professors:

Clark: "I don't want to ask anybody for help"

Professor: "Why not?"

Clark: "Because then I won't make it on my own."

Professor: "So what?"

Clark: "Then I won't deserve what I get."

Professor: "Why is that

Clark: "Because people who don't make it on their own just don't deserve anything
they get."

Professor: "Is that necessarily true?"

Clark: "Yes, people have to make it on their own."

15. A mother who is having difficulties getting her children to obey her says the following:

Mother: "All other things do what they're supposed to do. Birds fly south for the winter and the earth revolves around the sun. These things just do what they're supposed to do and without even being told. Why, then, don't my kids do what they're supposed to do?"

*16. Bert is feeling depressed. Here is what he tells his friend:

Bert: "Both of my cars are broken again. I swear, they must have planned this, just to torment me. Life is just closing in on me. I'm growing tired of fighting a losing battle. My job is a real bitch. The new house I bought just won't leave me alone. Everyday something else is the matter with it. I am sick and tired of being a slave to it. Things have just got me by the throat, and they just won't let go."

17. Mr. Presser, a high school math teacher, is sitting at his desk reading a book while his class works on an assignment. James, a student in the class, approaches Mr. Presser and inquires about the grade on a test that he had made up last week:

James: "Did you grade my make-up exam yet, Mr. Presser?

Mr. Presser: "I sure did."

James: "What grade did I get on it? Could you look in your grade book?"

Mr. Presser: "No, I don't have time, but don't worry, you did alright."

James: "But, Mr. Presser."

Mr. Presser: "Look, I told you that you did alright. Now go back to your seat and finish your assignment."

18. Lenny and Gerta are two high school freshmen in Mrs. Weiner's geography class. Gerta has been trying to transfer out of her class because she believes that she is not "learning anything" in this class. Here is a discussion be- tween Lenny and Gerta:

Lenny: "How do you like Mrs. Weiner's class so far?"

Gerta: "Oh, I'm just crazy about her class. She's a true scholar. They ought to place her name in the history books along with Albert Einstein and Socrates."

19. Here is a statement that appeared in a letter received by the Welfare Department in application for support:

"I want money as quick as I can get it. I have been in bed with the doctor for two weeks and he doesn't do me any good. If things don't improve, I will have to send for another doctor."

FURTHER PRACTICAL EXERCISES FOR CHAPTER FOUR

1. Dividing in smaller groups, let each group create and take turns *role playing* (before the whole class) at least one practice scenario containing at least one **Fallacy of Meaning**. The class may then critique the scenario, identifying the specific fallacy or fallacies committed and discussing how the characters in the scenario could have more effectively dealt with their situation. (*A Further Option*: After the class finishes its critique, the group may then "replay" the scenario, however this time omitting the previous fallacy or fallacies, dealing more effectively with the original situation.)

2. Collect at least 3 examples of at least 3 *different* **Fallacies of Meaning**. You may find these in newspaper editorials, magazine articles, advertisements, TV or radio commercials, political addresses, books, essays, etc. You may also use personal experiences you have had such as what you or somebody else said under a given set of circumstances. For each fallacy you find, write down the name of the fallacy and briefly describe why you believe it to be that fallacy. If you have *BELIEF-SCAN* you may use the program to support or establish your findings. Include in your project package copies of any supporting newspaper clips, advertisements, etc. that you have used (where available). Also attach any supporting *BELIEF-SCAN* reports. Be prepared to present and discuss your findings in class.

3. Select an essay or editorial (from any source you want) written on a controversial subject, and analyze it for **Fallacies of Meaning**. Write a brief evaluation of it's strengths and weaknesses in the light of your findings. If you have *BELIEF-SCAN,* you may use it to check for these fallacies.

CHAPTER FIVE
RELEVANCE FALLACIES

WHAT ARE RELEVANCE FALLACIES?

Relevance Fallacies occur when you try to support what you believe with statements which, although they may or may not be true, have little or nothing to do with what you believe. There are a number of different ways in which you can commit this kind of fallacy. These are discussed below.

1. APPEAL TO IGNORANCE

In the Appeal to Ignorance, you make the mistake of thinking that because something cannot be proven true, that it must therefore be false; or that because something cannot be proven false, that it must therefore be true. For example, Elsa says "If the authorities have not proven that my husband is a child molester, then my husband is not one"; and Justin says, "If you can't prove that John is not a drug addict then John probably is." Just because the authorities were unable to show that Elsa's husband was a child molester does not mean that he is not one. It only means that the authorities did not have enough information to prove it, one way or the other. Similarly, just because you can't show that John is not a drug addict does not prove that he is. In general, not having enough information to prove something true does not mean that it's false; and not having enough information to prove something false does not mean it's true. In order to prove something true or false you need to have "the facts." Ignorance of the facts is no defense!

2. APPEAL TO FORCE

The fallacy of Appeal to Force occurs when you use force instead of rational discussion. Instead of trying to give good reasons for what you believe or want, you instead threaten to do something unpleasant (for example, to do physical injury, to blackmail, to punish) if you don't get your way. For example, Arnold says, "If my son tells me he doesn't want to do his homework then I'll whip his butt"; and Stephen says "If my wife wants me to support her and the kids then she better have my dinner on the table when I get home." Although the use of force may sometimes be justified on some occasions (say, to defend yourself against a rapist) using force is itself irrational when you can use rational discussion as a way of trying to solve your disputes. This is so because, while the use of reason tends to open up avenues of communication and understanding, which are important conditions of conflict resolution, the use of force often tends to block them.

3. TWO WRONGS MAKE A RIGHT

The fallacy of Two Wrongs Make a Right occurs when you think that, because someone has wronged you, it is OK for you to wrong that person. For example, Humphrey thinks that, "If my wife cheated on me then it is alright for me to cheat on her"; and Howard says "If my friend said bad things about me behind my back, then its alright for me to do the same to him." In general, the old saw "Two wrongs don't make a right" is good logic: rather than righting a wrong, doing something else wrong simply adds to the number of wrongs in the world. This does not mean that you should not defend yourself against an attacker, for example. However, in such a case, your act of self-defense (as long as you do not use any more force than is really needed and you really need to use force) would not be wrong. The fallacy occurs when you respond with an act that you know is wrong, but you think that doing something wrong will somehow correct another wrong.

4. MISUSE OF AUTHORITY

The fallacy of Misuse of Authority occurs when you think that, because a person is an authority in one area, he or she must be an authority in some other different area. For example, Lloyd says, "Since my physician tells me that I don't have any psychological problems then I don't need to see any counselor." Of course, the physician, say an internist, may be an authority in internal medicine, but it is a fallacy to think that therefore the physician must also be an expert in counseling.

5. JUMPING ON THE BANDWAGON

The fallacy of Jumping On The Bandwagon occurs when you think that because others are doing something, that it is alright for you to do it also. For example, Mary, a teenager, says, "If my friends are drinking booze, then it's alright for me to drink booze." However, Mary needs some good reason for thinking that it's alright in the first place for her to drink alcohol. Since many people do many things without having any good reason for doing them, it is a mistake to think that it is alright to do something *simply because* others are doing it.

When you jump *blindly* onto a bandwagon, you lose control of where you may fall. While pressure for group conformity may often be high, the importance of "looking before you leap" should not be underestimated.

6. PARROTING

While Jumping On The Bandwagon involves *doing* what others are doing simply because they are doing them, Parroting involves *believing or say*ing something simply because others are believing or saying it. Insofar as to believe or say something is also *to do* something, Parroting may be conceived of as a kind of Jumping on the ("Thought") Bandwagon.

For example, Stan says "If people say that my girlfriend is a whore, then my girlfriend must really be one." Since what others say about something or someone may just be gossip, Stan needs further facts to support what is said. The mere fact that others are *saying* something is not itself good evidence for what is said. Therefore Stan cannot rightly accept that his girlfriend is a "whore" just because this is what has been said.

There is one possible exception to the above. If many *authorities* on a specific subject agree in an opinion on that same subject (the subject on which they are authorities) then it is not necessarily a fallacy for you to trust that opinion. For example, if several psychologists agree on the same diagnosis, then it may not be a fallacy for you to accept that diagnosis.

On the other hand, many instances abound in which general beliefs or rules--such as moral, religious, and social ones--are formed as a result of *a process of socialization.* Many of these beliefs or rules do not originate as a result of careful thinking or consideration of the testimony of experts. Rather, they arise from early childhood as a result of the *mere* acceptance ("internalization") of what others such as parents, teachers, and peers have repeatedly said or conveyed by example. For example, Mary has been brought up from early childhood to believe that women are inferior to men; and John has been brought up to believe that the man "should wear the pants".

However, unless John and Mary think carefully about their own socialized beliefs they may find themselves self-defeatingly parroting them back despite much frustra- tion. Thus, Mary may willingly accept an oppressive or abusive relationship; or she may deny herself the opportun- ity to pursue her own desired goals because her husband's should come first; and John may find it difficult or impossible to "make ends meet" because he thinks his wife should not also "wear the pants."

While socialized beliefs and rules may be very familiar and even *seem* to be self-evident (the Self-Evident Fallacy may also arise here), you still have the capacity to inspect these beliefs for their reasonableness. This capacity to think about and to understand your own thoughts and speech is, after all, what separates us humans from mere parrots!

7. APPEAL TO TRADITION

The Appeal to Tradition occurs when you think that, because something has been a tradition, it must always be done and must never be questioned. Although traditions can and do sometimes serve worthwhile purposes, this does not mean that they must always be respected *just because* they are traditions. For example, Larry says, "If my mother and my grandmother and my great grandmother stayed at home and raised children, then my wife must do the same." Larry's insistence on tradition for its own sake is a fallacy since it amounts to *blind* acceptance of tradition. Larry has closed his mind to any important differences between his wife's situation and those of his ancestors (for example, Larry and his wife are unable to live on Larry's income alone).

It is useful to keep straight the difference between the Appeal to Tradition and the fallacy of Insisting on the Past, which was discussed in Chapter 2. In the case of Appeal to Tradition, your mistake consists of thinking that, come what may, you have a *duty* to do what has been done in the past. In the case of Insisting On The Past, you do not see yourself as having a duty do carry the past into the present. Rather, you think that what has happened in the past will automatically happen in the future, regardless of what you do, so that there is no point in trying to change things. Both fallacies, however, typically lead to a similar self-defeating result, namely, failure to make constructive changes in your life!

8. MISUSE OF PITY

Although there is nothing in itself wrong with having pity for others, this emotion may sometimes be misused. In the fallacy of Misuse of Pity, you let pity influence your opinion instead of basing it on more appropriate considerations. For example, Liza says "If my father has a bad heart then they (the authorities) should not make my father have therapy for sexually abusing me." In such a case, it may be in order for Liza to feel sorry for her father's heart condition; but that pity should not be confused with a good reason for her father's not having therapy. In this case, the appropriate consideration for deciding upon whether Liza's father should have therapy is her father's *need* of therapy. Unfortunately, Liza's pity may have blinded her to this need of her father.

9. THE WORLD REVOLVES AROUND ME

This fallacy occurs when you think that, just because you believe, accept or want something, then that is the true reality, and everybody else must believe, accept or want it also. It is the mistake of supposing that *your* beliefs must constitute the ultimate standard or test of reality, as though anyone else's thoughts could not be acceptable unless they matched your own.

. For example, Ignatz says, "If I believe that premarital sex is wrong then my daughter should think so too"; and Marcia says, "If I say my children love their stepfather then my children *do* love their stepfather." What you *believe* is true is not necessarily what *is, in fact* true; similarly, what you *want* to be true is not necessarily what is, in fact true. The *mere* belief or desire for something to be true is not enough to make it true. In general, reality does not *have to* conform to your beliefs, although, to be true, your beliefs have to conform to reality.

Moreover just because you think your belief or desire is correct does not mean that everybody else has a *moral duty or obligation* to share that belief or desire. This extremely *intolerant* attitude is unrealistic as well as self-defeating for the purposes of having successful communication and relations with others, who are not apt to always perceive things as you do.

10. SWEEPING GENERALIZATION

A Sweeping Generalization occurs when you ignore exceptions to a rule. For example, Jessica, a minor, argues that she doesn't have to attend school if she doesn't want to because "America is a free country." Similarly, Herman, a manic, refuses to take his lithium because "All drugs are bad for you." Both Jessica and Herman are applying rules to cases to which these rules are not intended to be applied. The rule that "America is a free country" does not mean that you are free to do just *anything*--including not ever go to school. And the rule against taking drugs does not mean that people should never take any kind of medication such as that prescribed by a physician.

It is helpful to keep in mind that most rules, even very important ones, have exceptions. Thus, for example, even such an important value as free speech has its limits. While, in the United States, you are free to express your political views at a public forum, you do not have similar freedom to yell "Fire!" in a crowded auditorium. In general, part of what it means to know a rule is to comprehend the limits of its application.

Sweeping Generalization should not be confused with Hasty Generalizing (discussed in Chapter 2). The two fallacies are, in one respect, opposites. Whereas Hasty Generalizing involves a faulty process of *arriving at* a generalization, Sweeping Generalization involves a faulty process of *applying* one. In the first case, you generalize on the basis of an inadequate sampling of particular cases. In the second, you apply a generalization to particular cases to which it was not intended to apply.

11. PERSONAL ATTACK

The fallacy of Personal Attack occurs when you try to call into question what a person says or does by attacking the person rather than what the person says or does. For example, Jake says "If my father is an alcoholic then I don't have to listen to him when he tells me not to drink"; and Henry says, "If my sister's husband came from the ghetto, what kind of parent could he possibly be." An alcoholic who advises his child against drinking may not be "practicing what he preaches"; but his advice may still be good advice. Similarly, a person who was brought up in a ghetto may still be a good parent.

Note that this fallacy is committed *only if* the attack on the person in question is *irrelevant* to what that person is saying or doing. Thus, for example, if you are deciding whether or not to believe what a certain person says, and you have known this person to have often been dishonest in the past, then your bringing up this past dishonesty as a reason for questioning what this person is now claiming would *not* commit this fallacy, insofar as this evidence is deemed to be relevant to the truth of the person's claim.

Of course, it is not always cut and dried as to what really is to count as "relevant" and there may be room for disagreement. For example, in some states, the former unchastity of a rape complainant has been considered "admissible" (acceptable) evidence whereas in other states it has not. Should the fact that Mary was sexually promiscuous in the past be acceptable grounds for questioning the rape complaint she has made against Sam? Insofar as the answer to this question is no, bringing in this information would amount to a personal attack on Mary. Otherwise, it would not.

12. CHANGING THE SUBJECT

This fallacy is committed when you appeal to irrelevant statements, questions, commands, jokes, stories, etc. for purposes of avoiding the subject at hand.

For example, when Harry's teenage son asks Harry for the keys to the car, Harry asks his son how he has been doing in his Algebra class; and when George, a local politician, is asked by concerned citizens to justify the increase in property taxes, he responds by asking them if they would like to live under a communist government where there is no private property at all; and when

Ken asks his wife Kathy to have sex with him she talks about their dating days such as when the two of them were caught "making love" in the back seat of an old Chevy; and when Boris, a science teacher, is asked by a student to answer a question to which Boris himself does not know the answer, Boris responds in Russian and then asks "Does that answer your question?" The class then gets "lost in the laugh" and the question remains unanswered.

Since Changing The Subject can prevent important issues from being resolved or even addressed it typically tends to have self-defeating consequences. Thus, Harry's son is frustrated by not knowing if he can use the car; the citizens remain uninformed about the reasons for the tax increase; Ken is frustrated over his sexual prospects; and Boris' student is not even given a clue.

While there may be occasional circumstances where Changing The Subject may actually serve to divert attention from an issue that is best left unaddressed (for example, where discussing it would probably lead to needless quarrel), this is not usually the case. Rather, by avoiding Changing The Subject it is more likely that open and sincere interpersonal communication will take place, problems or disagreements will be resolved, and the common frustration of not even having your concerns taken seriously will be reduced.

13. HAIRSPLITTING

The fallacy of Hairsplitting is committed when you emphasize relatively insignificant details of an issue while failing to adequately address major or significant ones. Thus, like Changing The Subject, this fallacy leads you to neglect and divert attention from matters that would have better been addressed.

For example, when Mr. Milton, a high school social studies teacher, grades an essay exam, he spends most of his time finding misspellings and grammatical errors instead of looking at *what* the student has said; and when Seymour's son asks him for advice on dating, Seymour concentrates on the need for his son to have "clean white healthy teeth" and the "need to use a gel with fluoride"; and when Emily visits a doctor she talks about the doctor's physical appearance and age instead of considering the doctor's medical evaluation and advice; and when Willy and his wife eat out, he spends much of the meal time complaining to his wife about a dirty fork or about the quality of the service or about what he thinks is wrong with the food; and when Jack and his wife Abby go out to "have fun on a Saturday night," Abby spends most of the night out talking about what the two can do next week.

By losing sight of what is relatively significant and instead immersing yourself in less relevant issues of practical situations, you are likely to waste a lot of valuable time and to experience less happiness and greater amounts of frustration. For example, instead of enjoying *the present*, Abby wastes her night out wondering what to do later; but when the later becomes the present, she self-defeatingly does the same thing!

This fallacy may be avoided by making an effort to rank the *relative* significance of issues, that is the significance of an issue given the specific purposes at hand. This means that a given issue may or may not be significant depending upon the situation in question. For example, in the context of eating out, the most significant aspect may be the intimacy of dinning out together, and not your opinion that you have tasted better cuisine elsewhere. Consequently, *in this context and given these purposes* you would be hairsplitting by spending large amounts of time critiquing the food quality or service. On the other hand, if you are a restaurant reviewer writing a column for a newspaper, then the quality of the service or food may be of paramount importance. Similarly, given the usual purposes for visiting a doctor, a patient who concentrated on the physical qualities of the doctor rather than on his or her medical advice, would be hairsplitting.

While at least *some* cases of Hairsplitting may also involve Changing The Subject, these two fallacies are distinct. One major difference is that Changing The Subject (as discussed above) involves a *deliberate or intentional desire to divert attention* from the subject at hand. While hairsplitting may also involve diversion of attention from a significant issue to one of lesser significance, this is not necessarily the result of any intentional or deliberate attempt to divert attention. For example, while Seymour's advice to his son about successful dating is obviously hairsplitting (since given the purposes of dating there are more important things than having "clean white healthy teeth" and "the need to use a gel with fluoride"), the advice, although lame, may be given in earnest, not as a deliberate attempt to Change The Subject.

REVIEW DEFINITIONS: RELEVANCE FALLACIES

1. **APPEAL TO IGNORANCE** =*Df.* Believing something merely because one is not able to prove it false; or disbelieving something merely because one is not able to prove it true.

2. **APPEAL TO FORCE** =*Df.* Trying to persuade others by threatening them, even though one might have tried rational means instead.

3. **TWO WRONGS MAKE A RIGHT** =*Df.* Trying to do something to someone else that I think is wrong because I also think this person has done something wrong to me.

4. **MISUSE OF AUTHORITY** =*Df.* Taking the advice of someone because one thinks that this person has special knowledge, training or qualifications even though this person's qualifications are in some area that does not especially qualify him or her to give the sort of advice being given.

5. **JUMPING ON THE BANDWAGON** =*Df.* Doing something merely because one thinks others are doing the same thing.

6. **PARROTING** =*Df.* Saying or thinking something simply because you heard someone else say it.

7. **APPEAL TO TRADITION** =*Df.* Insisting on doing something now simply because it has been a custom to do it in the past, even though circumstances are different now and do not support the usefulness of continuing to do it.

8. **MISUSE OF PITY** =*Df.* Being emotionally manipulated by others into feeling sorry for them even though you know that there are good reasons for not giving in to their requests.

9. **THE WORLD REVOLVES AROUND ME** =*Df.* Thinking that, because you believe, want, or accept something, then others must do so too.

10. **SWEEPING GENERALIZATION** =*Df.* Trying to apply a general rule to some situation that the rule is not really supposed to be applied to in the first place.

11. **PERSONAL ATTACK** =*Df.* Trying to disprove something someone says or does merely by criticizing this person, even though the criticism has little or nothing to do with what the person is saying or doing.

12. **CHANGING THE SUBJECT** =*Df.* Trying to divert attention from the subject at hand by introducing some irrelevant line into the discussion.

13. **HAIRSPLITTING** =*Df.* Given the specific purposes at hand, focusing attention upon insignificant or irrelevant matters at the expense of addressing significant ones.

PRACTICE SCENARIOS FOR CHAPTER FIVE

EACH OF THE CASES PRESENTED BELOW CONTAIN AT LEAST ONE OF THE FALLACIES DISCUSSED IN THIS

CHAPTER. FOR EACH CASE, NAME AT LEAST ONE FALLACY IT CONTAINS AND EXPLAIN HOW THE PERSON WHO COMMITTED THE FALLACY MIGHT HAVE AVOIDED IT. THAT IS, EXPLAIN HOW THIS PERSON COULD HAVE MORE EFFECTIVELY DEALT WITH HIS OR HER SITUATION.

1. Hilda's father was, from the very start, against Hilda's marriage to Martin, and now, even after 10 years of being married to Martin, he still often urges Hilda to divorce Martin. "Hilda," he says, "you're still young and attractive and very bright. Why don't you find yourself someone who will amount to something. That Martin is, and always has been, nothing but a lazy lump." The following discussion occurs between Martin and Hilda:

Hilda: "Martin, you forgot to take the garbage out again."

Martin: "Oh darn it. Well there's still some room in one of the cans. I'll get it out for the next pick up. Sorry hon."

Hilda: "This isn't the first time you did this. You know what I think? I think you're nothing but a lazy lump who will never amount to anything."

2. Professor Atfield, a college English teacher, suspects that Jerry, one of his students, has plagiarized an assignment. In what follows, Mr. Atfield confronts Jerry on the matter:

Professor Atfield: "Jerry, this paper does not look like your work. It appears that it has been taken from another source although I admit that I haven't been able to pinpoint it. I believe, however, that it is not your paper."

Jerry: "You can't prove that I lifted the paper. You said yourself that you didn't even find the book I copied from. So, I rest my case. I didn't do it and you have some nerve accusing me."

3. Here is what one employer who runs a chain of grocery stores says just after he learns that one of his employees, who has worked for him at one of his stores for the past seven years, is gay:

"I don't want that queer working for me anymore. He's out of here. "

4. Becky has been divorced twice. Both of her husbands "played around." Brenda, who is Becky's close friend, has been married for six years. In the following discussion, Becky challenges Brenda's belief that her own husband has always been faithful to her:

Brenda: "My husband would never cheat on me."

Becky: "I wouldn't be so sure about that. How do you know that he never played around with some other women? You can't prove that he didn't, can you?"

Brenda: "Now that you put it that way, I can see how naive I've been. And to think that I trusted him all this time, that no good bum!"

*5. Sandy is a 20-year-old college sophomore. Here is a discussion she has with one of her professors:

Sandy: I'm really pretty stupid. I'll be lucky if I get a "C" in your class. I know it."

Professor: "Why are you telling yourself that? Your class discussion is excellent, and your essays show a lot of promise. I think that you are underestimating yourself."

Sandy: "Since I was a little girl, my parents have told me that I'm stupid. And they're my parents. They ought to know."

6. Winona's son Tommy is three years old. A working mother, Winona is looking for a day care facility in which to place her son. After speaking with some other parents, she comes to a conclusion. Here is her thinking:

Winona: "Several of the other parents I spoke to put their children in the Little Monkey Day Care Center. It must be good, then. I think I'll put Tommy in there too."

*7. Mort and his wife, Betty, have had a quarrel. The two are now not on speaking terms. Here is a discussion that Mort has with his older brother, Howard:

Howard: "Look, why don't you just make up with her. You are just making it difficult for the both of you."

Mort: "If I were you, Howard, I wouldn't go around giving other people advice about married life. You were divorced twice. How can you sit there and tell me how to run my marriage?"

8. Sharry has the following conversation with her counselor:

Sharry: "Last week my husband screamed at me just because I forgot to go grocery shopping. Can you imagine?"

Counselor: "How did you handle that?"

Sharry: "Since it happened, I've given him the silent treatment. Whenever he talks to me I just ignore him, like he's not even there."

9. Karen, a high school sophomore enjoys and has exceptional aptitude for foreign languages, especially French. Her French teacher has, in fact, encouraged her to consider pursuing French in college after she graduates high school. However, here is what Karen tells her teacher:

Karen: "My dad always tells me that I'm not really college material and that it is more important anyway for my brother to go to college. I guess I'll just keep my old job as a bagger at the supermarket when I graduate or maybe I'll go to secretarial school or something."

10. The following discussion occurs between Lenore and her husband, Chuck:

Lenore: "Dear, you've had that cough for over two weeks. Don't you think that you should see a doctor?"

Chuck: "OK, I'll go. Who should I go to? I don't know any doctors."

Lenore: "Why don't you call Dr. Preston. I heard she's real good."

Chuck: "I went to her once. She looks like a man. I can't take her serious. Do you know a doctor who knows their business?"

11. Claudine has just gotten a new evening dress, which she models for her husband, Claude. Here is their discussion:

Claudine: "Honey, how do I look in my new dress?"

Claude: "I think it might look better if you lost ten pounds. Your backside sticks out too much in it. You remind me of a freight train."

Claudine: "Thanks a lot. That's the last time I ask you for your opinion."

Claude: "Look, you asked me a question and I gave you an honest answer. A good marriage is based on honesty and that's just what I was, honest. I told you just what I was thinking."

12. "I'm sure I could do marriage counseling. After all, I've been married for twenty years."

13. Most of Mr. Murtle's high school geometry class is having problems understanding the current subject matter. Here is a discussion between Mr. Murtle and one student, Jeff, who is speaking on behalf of most of the other students in the class:

Jeff: "Mr. Murtle, can we spend more time reviewing this material before starting a new unit? Most of us really still don't know what's going on?"

Mr. Murtle: "I'm sorry, but there are requirements to satisfy. I can't take the time to go over it again. I need
to go on to other things so that I cover everything I'm supposed to

cover."

*14. In what follows, Terry, a 12-year-old girl is trying to convince her father to allow her to see an R-rated movie:

Father: "You cannot see that movie, and that's all there is to it. It is rated "R" and you are only twelve years old."

Terry: "You treat me like such a baby. Why don't you ever trust me. Don't you even care about my feelings?"

[She starts to cry.]

Father: "OK, OK, stop that crying. You win. You can see the movie."

Terry: "Thanks Dad. I love you. You're the greatest."

15. Susan recently separated from her husband, Alan, and has filed for a divorce. Alan, a crack addict, had spent almost all of the couple's savings on his drug habit. Whenever Susan asked Alan to get help, he simply denied he had any drug problem. Finally, Susan left him. On one occasion after their separation Alan visited Susan for purposes of asking her for money to buy drugs. Alan, at this visit, showed clear signs of withdrawal; he was poorly groomed and had not changed his clothes in weeks; he was in tears. Their discussion was as follows:

Alan: "I lost everything--my job, the car, you. I have nothing left. Please Suzy baby; I'm still your husband. I need the money bad, baby. I'm in big trouble. Please help; please help me, Suzy; please."

Susan: "OK, I'll give you what I have. How can I say no when you look so pathetic."

16. Hubert, a freshman in high school, has a book report to write in English. His friend Jorge has recommended that he do it on "The Catcher in the Rye." Here is what he tells Jorge:

Hubert: "That book you recommended stunk. It had a boring

red cover with no pictures and just the title and author written on it. So I decided to choose a different book."

17. Jill is 17 years old. Recently her father was tried and found guilty of the sale of illegal drugs. Despite the conclusive evidence against him, Jill insists upon her father's innocence. The following is a discussion between Jill and her older brother, Ted:

Jill: "Dad is innocent. He would never have done anything like that. You must believe that too."

Ted: "But, Jilly, they found him guilty; there was no reasonable doubt."

Jill: "I don't care. I still think he's innocent, and that's all that really counts."

18. Bobby wants his 12-year-old stepson, Eric, to learn how to shoot a rifle. Eric, however, doesn't want to learn and Eric's mother thinks that Bobby should respect Eric's wishes. But Bobby finds this hard to accept. Unlike his wife and stepson (who are from the city), Bobby was born and raised in the country where he learned to shoot at a young age. Hunting for food was, for him, a large part of his childhood. Here is what Bobby tells his wife:

Bobby: "My father taught me to shoot when I was a boy. This is how it's always been in my family; and this is how it's always gonna be. Whether he likes it or not, the boy's gonna learn to shoot."

19. Amanda has been dating Pete for four months. Recently, Pete has begun to talk about getting married. Amanda, however, does not want the relationship to go any further although, in spite of her feelings, she has continued to date him. Here is what she tells a close friend:

Amanda: "I really want to break it off, but I just don't know how to tell him. The last time I tried, he looked like he was going to cry. He's a nice guy and all, and I really don't want to hurt him."

20. Melanie's 10-year-old son, Mike, has asked his mom, on many occasions, for an ATV (All Terrain Vehicle). Until recently, Melanie has consistently refused this request on the grounds that they are very unsafe. However, here is a recent discussion between the two in which Melanie changes her mind:

Mike: "All the other guys have em, mom. I'm the only one who doesn't and I look like a nerd. Oh please can I have one? I'll be careful. Please mom?"

Melanie: "Well, alright but you be careful, you hear?"

21. Denise and her boyfriend, Wayne, are planning their weekend activity together. Here is their discussion:

Wayne: "Let's go fishing at Cricket Creek. That's real fun."

Denise: "But I'd like to go swimming at the country club. I don't really like fishing."

Wayne: "Boloney. That country club stinks. Fishing at The Creek is better. Besides, I know you'll have fun. I always do; so you will too."

22. Here is what Stanley tells a friend about his blind date last night with Fran:

Stanley: "I had a great time with Francine last night. We really seemed to hit it off. She's very nice, and I really enjoy being with her. She's really good looking too except there's one thing that bothers me about her. She's got a bunion on her big toe and I really don't want a girlfriend with ugly feet."

23. Darien is a high school senior who is enrolled in Mrs. Drucker's chemistry class. He is presently in danger of failing chemistry and Mrs. Drucker has discussed the matter with Darien's parents. Mrs. Drucker has agreed to provide additional help for

Darien after school and Darien has agreed to delegate more time to studying chemistry. In fact, he has decided to spend his Saturday afternoon's doing his chemistry instead of socializing. Here is a discussion be- tween Darien and Mrs. Drucker at the end of one of their extra help sessions:

Mrs. Drucker: "Darien, I need someone this Saturday to help me move into my new home. You have a strong back. Would you like to help?"

Darien: "Well, I don't know; I was planning. . . . Well, alright, I think I can. OK, I can do it."

Mrs. Drucker: "Are you sure? I can find someone else."

Darien: "Sure, no sweat!"

24. Fenton and his wife, Juliet, have both come home late from work. Neither has eaten supper yet and Fenton suggests that they go out to eat. Here is their discussion:

Fenton: "Why don't we eat out tonight?"

Juliet: "Good idea. Where do you want to eat?"

Fenton: "I could go for some Mexican food. Let's go to Los Sombreros. That's a good place."

Juliet: "I really can't eat that stuff. I get sick from it. Can't we go somewhere where I can get something I can eat too?"

Fenton: "Will you stop being so disagreeable? Mexican food is good food. It never makes me sick; and it tastes good. Come on, let's go. It's getting late and I want to get back in time so we can watch that TV show I told you I wanted to see."

25. Leo and Tabatha have been married for twenty years. Much of their time together has been spent quarreling with one another. In what follows they are both dressing for an affair that they will be attending together later on that evening:

h

Tabatha: "Why don't you wear your white jacket? And put on some deodorant. You smell."

Leo: "How would you know anything about fashion? Are you some kind of beauty queen? You fat pig."

Tabatha: "OK, that about did it. You think you can talk to me like that and get away with it? I'll get you back for that, you baldheaded little man. You don't even know how to make love. Not only that, you're blind as a bat without those ugly little glasses of yours. And you're boring too. Do you think I like listening to those stupid stories you tell me about that dumb little job of yours. And You think you're such a gem? Ha!"

26. John has recently lost his job as a shoe salesman and has been unsuccessful so far in finding another job selling shoes. Here is what John's wife tells her friend:

"John had four job interviews this month and he still hasn't gotten a job. How can I expect him to go through another rejection; so, if he wants to stay home from now on, I guess I'll have to take on an extra job in order to make ends meet."

27. In the lunchroom, Brandon, a very popular high school student is aggressively "mocking out" Elmer, a much less popular student, while several other students watch with amusement. For instance, Brandon calls the somewhat overweight Elmer a "Goodyear Blimp." Elmer appeals to the cafeteria monitor, Mr. Max, who is onlooking, to intervene and cause a halt to Brandon's verbal abuse. Here is How Mr. Max handles the situation:

Mr. Max: "Now Elmer, maybe you shouldn't be so thin-skinned. Then again, that might be the only thing `thin' about you."

[The other students laugh]

28. "Nobody in my family has ever gotten a divorce so I'll be damned if I will!"

29. Here is a discussion between Talbot and his 8-year-old son, Adam:

Adam: "Dad, do you believe in God?"
Talbot: "Of course I do, son."
Adam: "But how do you know He exists?"
Talbot: "You just have to have faith, son."
Adam: "But that doesn't prove anything. Why should I have faith anyway?"
Talbot: "Son, it's past your bedtime any I have things to do. Why don't you take a glass of milk and get to bed. Did you brush your teeth yet?"

30. "Since I should be making my own decisions, I don't want to hear your advice or anybody else's for that matter."

31. The following is a discussion between two high school students about one of their teachers, Mr. Roswell:

Bill: "Mr. Roswell is always lecturing us about how dangerous it is to be promiscuous in this age of venereal diseases, especially AIDS. Well guess what?"
Phil: "What? Tell me."
Bill: "My older brother went to college with Mr. Roswell and said that the man was "a real babe killer."
Phil: "So what does he expect us to do, stay virgins while he makes it with the women."
Bill: "I don't believe a guy like that. What a hypocrite!"

32. Jim: "Why did you loan my videos out without even asking me?"
Jim's Roommate: "Well, you take my compact disks without asking my permission, don't you? So I guess that makes us even."

*33. Maureen is single and extremely wealthy. She and Jim have been going together for three weeks when Jim asked Maureen to marry him. Here is a discussion between Maureen and her best friend Milley:

Maureen: "Jim proposed to me last night."

Milley: He did? How do you know that he's not just after your money? You don't know him very long, so you really don't know if he's on the level. So, I wouldn't believe him. He's probably just one of those con artists."

Maureen: Maybe you're right. I better tell him that I'm not interested in marriage. Maybe I shouldn't even see him anymore, not if he's some kind of con artist. He's very charming, but I really don't know what his real motives are. I guess I just better break it off now before it goes any further."

34. Morty has high blood pressure and is under a doctor's care. His wife usually comes with him when he sees the doctor. After one visit, here is what Morty tells his wife:

Morty: "You ask the doctor too many questions. Nobody else I know asks as many questions as you. They just shut up and do whatever he says, and that's what you should do too."

35. In what follows, Billy, a 7-year-old, asks his father if he can use some of his tools. Here is their discussion:

Billy: "Dad, can I use your electric drill to make an air hole in my toy chest so that I can put Ralphy [the dog] in there?"

Dad: "No. You keep your little hands off of my tools."

Billy: "But dad, why can't I?"

Dad: "Did you hear what I said!? If you touch any of my tools, I'll beat the daylights out of you, that's why! You keep your hands off of those tools or else! Do you understand?"

Billy: "Yes Dad."

*36. Preston Porterhouse, M.D. wants his son, Tom, to be a premed major in college, however Tom does not want to be a physician. Here is their discussion:

Preston: I am a physician and so was your grandfather and

your great grandfather. And I expect you to follow in these same footsteps, son."

Tom: "I don't know Dad. I Just don't think I'm cut out for it. I would rather major in music. I'm really good, Dad."

Preston: "Nonsense. You're a Porterhouse, aren't you?"

Tom: "Yes, Dad."

Preston: "Then that settles it, son. I don't want to hear any more of this talk about music."

37. Billy and Connie are two high school seniors who have gone steady for the past two months. Billy wants to have intercourse with Connie but, so far, Connie has refused. Here is what Billy finally tells Connie:

Billy: "If you won't go all the way with me, then I'm breaking up with you."

38. Gloria and her friend Kara have agreed to clothes shop for the afternoon. Here is their discussion:

Kara: "Why don't we look for some things at Good Deals Department Store. They're having a big sale today."

Gloria: "Are you serious!? Really, how could you even suggest such a thing! I would never buy my clothes at Good Deals. People of our class don't shop there. They buy their cloths in the Meadow Lark shops. That's where I suggest we shop, shall we?"

*39. Here is a discussion between two friends, Pam and Nora:

Pam: "I've been having these dizzy spells, off and on, lately."

Nora: "Don't you think you should see a doctor?"

Pam: "No. I'll be alright. I just have to get more rest. I talked to my friend Sharon yesterday and told her about the dizzy spells. You know she's Dr. Spencer's receptionist and she sees a lot of sick people. Well, she told me that I'm probably just tired and need to get more sleep. I think she's probably right; I really haven't been sleeping very well lately. I sure felt relieved after talking to her. I was imagining all sorts of things at first."

40. Here is what one English teacher says in response to a proposal to introduce new computer software for teaching English grammar:

"When I was a student, we didn't have any of this high tech equipment. We didn't have spell checkers, word processors, computers, mini calculators, and all these software packages. We didn't have it then and we still learned. So why do we need it now."

41. Daughter: "I'm not sure I ever want any children."
Mother: "Nonsense! The happiest moment of my life was when I gave birth. You'll feel that way too when you have children."

42. Here is a discussion between Fred, a community college student, and his religion instructor, Mr. Cornwallis:

Fred: Mr. Cornwallis, why did I get a "C" on my paper on "gay rights'?

Mr. Cornwallis: "Because you said that homosexuality is an acceptable sexual orientation. But quite to the contrary, homosexuality is a sin against nature. It is a sickness, not an acceptable life style at all, and society should not condone gay rights."

Fred: That's just your opinion. I read that not even the American Psychological Association any longer considers homosexuality to be sexually deviant behavior.

Mr. Cornwallis: "No, it's not just my opinion, its in the Bible. I don't care what the psychologists say. They apparently don't read the Bible, but I do."

43. Lynn is a single parent. Her daughter, Priscilla, an eighth grader, has displayed aggressive behavior, especially in school. In what follows, Lynn discusses this matter with her girlfriend, Lisa:

Lynn: "Priscilla has been getting in trouble in school. They say she has been cutting classes, disobeying school rules and speaking disrespectfully, using profane language, to her teachers."

Lisa: "But Priscilla was always so good in school. What are you going to do about it?"

Lynn: "They want Priscilla to see the school psychologist, but I don't think that's necessary."

Lisa: "Why not?"

Lynn: "I told my father about Priscilla, and he says that it's just a normal stage that adolescents go through. He ought to know. Right? He raised me."

44. Jackie wants to have dinner at her mother's next Sunday. She tries to convince her husband Ed to go with her but Ed refuses. Here is what he says to Jackie about the matter:

Ed: "You know I have a heart condition, so you better not argue with me.You wouldn't want to be responsible for my dropping dead, would you?"

45. John and Cindy have been married for twenty-one years. In the past few years, the two have had sex on the average of once each month:

John: "Hey honey, you look good tonight. How about you and me? Ya know what I mean babe?"

Cindy: "Well aren't we frisky tonight. What would you like for supper? I'll tell you what. I'll make you those nice turkey burgers you like."

FURTHER PRACTICAL EXERCISES FOR CHAPTER FIVE

1. Dividing in smaller groups, let each group create and take turns *role playing* (before the whole class) at least one practice scenario containing at least one **Fallacy of Relevance**. The class may then critique the scenario, identifying the specific fallacy or fallacies committed and discussing how the characters in the scenario could have more effectively dealt with their situation. (*A Further Option*: After the class finishes its critique, the group may then "replay" the scenario, however this time omitting the previous fallacy or fallacies, dealing more effectively with the original situation.)

2. Collect at least 4 examples of at least 4 *different Fallacies* **of Relevance**. You may find these in newspaper editorials, magazine articles, advertisements, TV or radio commercials, political addresses, books, essays, etc. You may also use personal experiences you have had such as what you or somebody else said under a given set of circumstances. For each fallacy you find, write down the name of the fallacy and briefly describe why you believe it to be that fallacy. If you have *BELIEF-SCAN* you may use the program to support or establish your findings. Include in your project package copies of any supporting newspaper clips, advertisements, etc. that you have used (where available). Also attach any supporting *BELIEF-SCAN* reports. Be prepared to present and discuss your findings in class.

3. Select an essay or editorial (from any source you want) written on a controversial subject, and analyze it for **Fallacies of Relevance**. Write a brief evaluation of it's strengths and weaknesses in the light of your findings. If you have *BELIEF-SCAN,* you may use it to check for these fallacies.

CHAPTER SIX
FALLACIES OF MAKING VALUE JUDGMENTS

WHAT ARE FALLACIES OF MAKING VALUE JUDGMENTS?

This type of fallacy is committed when you misuse evaluative words such as (for example) "must," "ought," "should," "awful," and "terrific." Fallacies of this kind are discussed below. As you will see (in Chapter 8) this type of fallacies is often behind intense, self-defeating emotions.

1. DEMANDING PERFECTION (IN THIS IMPERFECT WORLD)

This fallacy occurs when you demand that some state of perfection must or has to be. Since the world is not a perfect place, such demands can only be unrealistic. For example, Donald falls into this mistake when he thinks that everyone must always like and approve of him. Since it is unrealistic for anyone to expect approval from everybody all of the time, such a demand is unreasonable. Some other common forms of belief that contain this fallacy are: "I must never fail at anything," "People must never do anything to hurt me," "Everyone must always treat me fairly," "I must never suffer or be inconvenienced," "Things always have to go my way," "I must always be in control of things," "I must always have the approval of others," "I need to be loved/respected by those people in my life who are important to me."

Notice that these beliefs use words of necessity like "must," "need," and "have to." When you use these words as they are used in the above beliefs you are making *demands*. Since these demands are for states of *perfection,* they are irrational demands.

It is at least sometimes *desirable* for others to approve of you, to be loved and respected by others, for things to go your way, to be in control, and so forth; but there are simply no laws of nature (like the law of gravity) that make these things necessary. Moreover, it is not possible to bring such laws into existence by demanding them. A more rational philosophy of nature is to accept that the world (and this includes all human beings in it) is imperfect and fallible.

Since you, as a human being, are fallible, it is also irrational for you to expect that you never commit any of the fallacies discussed in this book. No matter how good you may become at avoiding fallacies, you are still sometimes going to slip up. Remember that it is also a fallacy to demand perfect rationality!

2. I-CAN'T-STAND-IT-ITIS

This fallacy occurs when you use strong emotional language to exaggerate your ability to stand up to some problem in your life. For example, Todd claims that, "If my girlfriend broke up with me then I couldn't stand it"; Eugene says "I just can't tolerate my wife anymore"; Hilda says "If I lost my job then I just couldn't take it, I just couldn't live with it."

While some things may be justly called "unbearable" or "intolerable," such words are wrongly used in cases where life still holds out reasonable hopes for the future, in spite of the loss of something of value. For example, some levels of physical pain with no reasonable chance of relief may be justly called "unbearable"; but the same is not likely to be true in cases of a failed romance or a lost job.

It is sometimes easier to avoid a difficult, painful, risky or potentially frustrating task or problem than to face it. Sometimes, by telling yourself that the task or problem is so difficult, risky, painful, or potentially frustrating that you just couldn't stand it, you may take the "easier" way out. However, what is easier or less frustrating in the short run may sometimes prove to be more difficult, more frustrating, or at least less gratifying in the long run. For example, Seymour refuses to ask Jane out on a date, even though he would like to go out with her, because he couldn't stand it if she rejected him. In Seymour's case, it may be easier to avoid asking Jane out since he does not then risk the frustration of having her refuse his invitation. However, the avoidance of this possible frustration may be purchased at the expense of future happiness.

Seymour has also exaggerated the results of being rejected by Jane. Is it really true that Seymour *couldn't* stand it? Indeed, there is a difference between choosing not to stand something and *not being able* to stand it. Thus, Seymour could not stand the force of a Mack truck striking him at 100 miles per hour; but he *could* (choose to) stand Jane's rejection. There is also a difference between saying that something would be difficult or frustrating and saying that you just couldn't stand it. While Seymour might find Jane's rejection frustrating, this is, again, not to say that he couldn't stand it.

3. AWFULIZING

This fallacy occurs when you use strong emotional language to exaggerate the negative side of situations that you would prefer not to be in. For example, Sal says "If my wife divorces me then it would be horrible"; Sylvia says "It is terrible that I cannot bear any children"; Lawton says "It was awful that I failed my law boards." While some things such as the gas chamber and the guillotine may be "awful," "horrible," and "terrible," such words are misused when they are used in cases where reasonable hopes for the future still exist. For example, Sylvia could consider adoption or find some other outlet for her affections; and Lawton could take the exam again or consider a different line of work.

4. SLIPPERY SLOPE- AWFULIZING-I-CAN'T-STAND-IT-ITIS SYNDROME

Awfulizing often accompanies the fallacy of Slippery Slope. In such cases, first you Slippery Slope and then you Awfulize about it. For example, Jim tells himself that, because he made a mistake at work today, he is going to lose his job. Then he tells himself how awful, horrible and terrible this will be. Moreover, I-Can't-Stand-It-Itis (discussed above) also frequently occurs in such contexts. For example, Jim may go one step further and conclude that, because losing his job is so awful, he just cannot stand to lose his job. Indeed, since Slippery Slope, Awfulizing and I-Can't-Stand-It-It is so often occur together in this identifiable pattern or "syndrome," it makes sense to look for Awfulizing and/or I-Can't-Stand-It-Itis if you catch yourself sliding down the slippery slope! (This *Slippery Slope-Awfulizing-I-Can't-Stand-It-Itis Syndrome* will be discussed further in Chapter 8 regarding stress management.)

5. DAMNATION

This fallacy occurs when you negatively rate your *entire* self (*Self-Damnation*) or that of another person (*Damnation of Others*) because of something about yourself or the other person that you do not like. For example, Jenny says "If my husband lied to me then he's no damn good"; Lester says "If my father is an alcoholic then he's nothing but a worthless bum"; and you may also damn yourself as when you tell yourself that "If I lost my job then I am a complete failure in life." However, every person, yourself included, has many different sides of his or her personality. It is therefore a mistake to think that, because a person has some negative aspects, then the whole person must be damned.

Self-Damnation may be committed by making your self-worth dependent on such things as the love or approval of others

(*Approval Damnation*) and on your personal achievements (*Achievement Damnation*). For example, Tanya thinks that she is a "no good" person because her parents never really approved of her; and Andy thinks that he is "worthless" because he has not done well in school. However, lack of parental approval does not make Tanya "no good"; and poor grades do not make Andy "worthless." In calling themselves "no good" and "worthless" Tanya and Andy have denied *any* present and future value to their existences. Tanya has denied that there is, or can *ever* be, *anything* good about her because her parents do not approve of her; and Andy has similarly denied that he does, or can ever do, anything worthy because he has done poorly in school. But these are clearly irrational conclusions for anybody to draw about themselves.

Instead of damning your entire being on the basis of your behavior, it is more reasonable to rate your *behavior* instead, leaving your *self* out of the rating. For example, Andy can admit that he has not done well in school without damning his entire person. By separating his self-worth from his performance in school, Andy is more apt to deal constructively with his difficulty in school, for example, by looking for ways of improving his grades. He may also avoid self-defeating emotions and behavior ("Why should a worthless so-and-so like me even try anyway!")

6. POISONING THE WELL

The fallacy of Poisoning The Well occurs when you use strong negative emotional language in order to make something sound undesirable. Brenda says that, "If my husband wastes our vacation visiting his mother then my husband is nothing but a momma's boy tied to his mother's apron strings." In this case, Brenda may not wish to visit her mother-in-law, so she tries (whether on a conscious level or not) to "poison the well" against it. Since Brenda's husband probably does not want to be viewed as a "momma's boy tied to momma's apron strings," he may be quietly "seduced" by these words into doing just what Brenda wants him to do. The trouble with this way of relating, however, is that

Brenda and her husband have not communicated their true feelings to each other. By playing such psychological "games," the lines of honest communication have been blocked for this couple.

7. THE "TERRIFIC" FALLACY

This fallacy is the opposite of Awfulizing (which is discussed above). It is committed when you use strong positive emotional language (words of praise such as "terrific," "perfect," and "the best") to exaggerate the virtues of something or somebody. For example, Mindy says that, "If my mother and father get back together then everything would be perfect"; Kathy says, "If you knew my husband then you would see that he is the nicest guy in the world." Indeed, things are rarely if ever "perfect"; and "the nicest guy in the world" does not really exist. Words like "terrific" and "perfect" may in fact be used to avoid con- fronting important realities. For example, Mindy's mother and father may not have worked through their marital problems thereby making their reunion a less than perfect situation; and Kathy may be in a state of denial about threatening facts about her husband (for example, that her husband is having an affair).

8. THE "I JUST CAN'T HELP THIS FEELING" FALLACY

This fallacy occurs when you deny any responsibility for your own feelings, blaming them instead on events that you can't control. However, even though many events are not in your power to control, the way in which you evaluate and interpret those events is normally in your control. What upsets you are the things that you tell yourself about those events and not those events alone. Of course, had the event in question not happened, you might not have gotten upset. Nevertheless, your negative *value judgment* of what has happened is a necessary ingredient; and this ingredient is one that you normally *can* control.

For example, Trisha commits the "I just can't help this feeling" fallacy when she tells herself that her husband's failure to remember her birthday was what made her so upset and that she just couldn't help feeling deeply hurt by this. Similarly, Jack commits this fallacy when he says "My wife got me ripping mad when she called me a fat slob. She really did it then; I just exploded! And, so I hauled off and hit her." However, Trisha has actually *made herself* upset over her husband's forgetfulness by telling herself how terrible, awful and horrible this was; and Jack has *made himself* "ripping mad" by telling himself how terrible, awful and horrible it was to be called a "fat slob" by his wife, and what a rotten person she was for saying so. Indeed, more rational evaluations by Trisha and Jack would have allowed them to avoid self-defeating emotions and behavior. For example, Trisha *could* have told herself, "How disappoint- ing that he forgot my birthday, but it isn't really so terrible after all"; and Jack could have told himself, "I wish she hadn't said that, but this doesn't mean she's a rotten person who has just done something terrible to me."

Among the expressions that people use to hide their responsibility for their feelings are: "You (it) got me (made me) upset (angry, depressed, etc.)," "I got upset because of that," and "I can't help being upset by that." It is more accurate for you to say, "I got myself upset over that" or "I have allowed myself to be upset." By speaking in this more accurate way, and by catching yourself when you speak in the other way, you may be less likely to fall into the fallacy in question.

9. THE "THOU SHALT UPSET YOURSELF" FALLACY

This fallacy occurs when you think that you have some sort of duty to constantly make yourself miserable, de- pressed, or anxious whenever you perceive some actual or possible event in your life as threatening. For example, Jim thinks that he must preoccupy himself with, and make himself miserable, about a problem he is facing at work, and that he must stay in this state of gloom just as long as the problem persists. And Jane thinks she must remain in a state of anxiety until she brings her failing grades up to passing ones. However, life is often filled with difficulties and just as soon as one disappears it is not unlikely for another one to crop up. Jim and Jane will therefore find few occasions for happiness as long as they commit this fallacy. More- over, imposing upon yourself a "duty" to be miserable is not likely to solve your problems. A much more profitable approach is to come up with a reasonable strategy for solving your problem and then to act upon it (without, however, telling yourself that it must be a *perfect* strategy and that you must remain preoccupied and miserable until you get one.) For example, Jim could decide to speak with his boss and Jane could change her study habits; but even if these things don't work, there is no need, no special duty, no commandment, and no law of nature telling you to be miserable until matters improve.

REVIEW DEFINITIONS: FALLACIES OF MAKING VALUE JUDGMENTS

1. **DEMANDING PERFECTION (IN THIS IMPERFECT WORLD)** =*Df.* Insisting upon or requiring that the world or some part of it exist without any defects or flaws.

2. **I CAN'T-STAND-IT-ITIS** =*Df.* Underrating your ability to endure a situation that you perceive to be difficult or frustrating.

3. **AWFULIZING** =*Df.* Exaggerating just how bad a situation is by rating it as the worst or nearly the worst thing possible.

4. **SLIPPERY-SLOPE-AWFULIZING-I-CAN'T-STAND-IT-ITIS SYNDROME** =*Df.* First, exaggerating the negative (bad) effects of an action or event; second, thinking these effects to be awful; and third, thinking them too awful to stand.

5. **DAMNATION** =*Df.* Negatively rating your entire self (*self-damnation*) or that of another person (*damnation of others*).

6. **POISONING THE WELL** =*Df.* Trying to intimidate or manipulate another into doing what you want through the use of name calling or other similar verbal tactics.

7. **THE "TERRIFIC" FALLACY** =*Df.* Exaggerating the virtue of an individual (person, thing, etc.) by rating it as perfect or near perfect.

8. **THE "I JUST CAN'T HELP THIS FEELING" FALLACY** =*Df.* Blaming your emotions entirely on external events without acknowledging the role of your own thinking in causing those emotions.

9. **THE "THOU SHALT UPSET YOURSELF" FALLACY** =*Df.* Telling yourself that you must (have a special *duty* to) preoccupy and make yourself miserable over a perceived problem until it is resolved.

PRACTICE SCENARIOS FOR CHAPTER SIX

EACH OF THE CASES PRESENTED BELOW CONTAIN AT LEAST ONE OF THE FALLACIES DISCUSSED IN THIS CHAPTER. FOR EACH CASE, NAME AT LEAST ONE FALLACY IT CONTAINS AND EXPLAIN HOW THE PERSON WHO COMMITTED THE FALLACY MIGHT HAVE AVOIDED IT. THAT IS, EXPLAIN HOW THIS PERSON COULD HAVE MORE EFFECTIVELY DEALT WITH HIS OR HER SITUATION.

1. Jerry lost his job 10 months ago and although he has had a number of interviews, he has not since been able to get another job in his field. In what follows he discusses the matter with his counselor:

Counselor: "How are you feeling today, Jerry."

Jerry: "I have a job interview tomorrow. I really need this job and it's just right for me too."

Counselor: "What if you don't get it?"

Jerry: "That would be terrible."

Counselor: "Terrible?"

Jerry: "Yes, I just couldn't take another rejection. I don't know what I'd do then."

2. Here is what one teenager tells her boyfriend:

"You have the perfect family; I don't think your parents ever even fight; not like my family."

3. Here is what one junior high school teacher tells a colleague:

"Some of those little juvenile delinquents are really getting on my nerves. The other day one of them brought a dead mouse to school and put it in my desk. Today one of them started whistling Dixie while I was teaching. Not all of them are irksome but others really aggravate me. I'm becoming a nervous wreck from them. I just can't stand much more of it before I explode. Sometimes, I feel like shooting them."

4. Housewife: "I can't believe I couldn't even remember to turn off the oven. I couldn't even do that right. What good am I anyway!"

5. Henry is a newlywed. His wife, Diana, wants Henry to help with the household chores--grocery shopping, cooking and cleaning. Henry discusses the matter with his friend, Jim, who used to date Diana before she met Henry. Here is their discussion:

Henry: "Diana wants me to split the grocery shopping, cooking and cleaning with her."

Jim: "Are you serious? You mean she's going to try to get you to do her work too? Boy, if you stand for that then she'll turn you into a henpecked wimp before long."

Henry: "No she won't; because I'm not going to do it."

Jim: "That's the spirit. Now you're talking like a man."

6. Dr. Molley, a university instructor, stops in the midst of one of his lectures and takes half of the class hour to address one student who had his head down on his desk apparently sleeping instead of taking notes. Here is some of what he says:

Dr. Molley: "You really have gaul thinking that you can sleep in my class. This is one thing that I just cannot, and will not, tolerate in my class. You are setting a bad precedent for others to follow. If you got away with it, then others would try it too and eventually the whole class might follow your example. You could destroy the whole point of being here! This would be an awful thing, and I just couldn't stand to let you undermine my class like that."

*7. Lucille's husband, Ted, had an affair with another man while they were married and subsequently left here for this man. Here is a discussion between Lucille and her counselor:

Lucille: "It's bad enough when your husband leaves you for another woman, but its a terrible thing when he leave you for another man."

Counselor: "Why is it a terrible thing?"

Lucille: "It means that I couldn't even compete with another man. What kind of woman am I if I can't even do that."

8. David's mother has invited him and his wife, Stacey, for dinner next Sunday. Stacey is opposed to the idea of going. Here is what she tells David:

Stacey: "Do we have to waste another Sunday evening listening to her? I just can't stand the way she treats me. She hardly ever talks to me and, when she does, she usually tells me how wrong I am. She makes me feel like such a jerk. Every time I leave her house I can't stop shaking. She makes me so upset, honey. Do we really have to go? Can't we tell her that our car is broken or the kids have chicken pox? That way she won't be able to aggravate me and destroy another Sunday evening for me."

9. Below is a discussion between a professor and one of his students, Hank, who has just admitted to cheating on all of his exams:

Professor: "Why have you cheated on your exams?"
Hank: "Because then I don't have to try to do good."
Professor: "What's wrong with trying to do good?"
Hank: "Then I might fail."
Professor: "And what if you try to do good and fail?"
Hank: "If I try to do good and fail, then I'm a failure, I'm worthless."
Professor: "Do you think you should never fail, and if you do, then you're a failure?"
Hank: "No. It doesn't bother me if I don't try to do good and fail. I have to be successful at what I try; otherwise I'm a failure."

*10. Roselyn is a 30-year-old housewife with two young children, ages 6 and 8. Much of her daily routine consists of doing household chores, transporting children to and from school, preparing meals, helping her children with their homework and a number of other domestic activities. Roselyn often schedules conferences with her children's teachers to discuss her children's progress in school. In a recent conference, Roselyn was told that her son, Bobby, has been getting into fights with some of the other children at school and has been talking in class during rest period. Here is a discussion between Roselyn and her husband, Ross:

Ross: "Honey, why don't we go out dancing this Saturday."
Roselyn: "I don't think so; not this Saturday; maybe next

weekend."

Ross: "Why not? We'll have some fun."

Roselyn: "We don't even have a sitter for the kids."

Ross: "I spoke to Jackie next door and she said that she would be happy to sit."

Roselyn: "Is that all you can think of at a time like this? Doesn't it even bother you that your son is having problems in school?"

Ross: "Sure it does Honey but why can't we go dancing anyway? Don't worry, I'll get to the bottom of this. Just don't you worry honey!"

Roselyn: "Look, Ross, leave me alone this weekend. I'm just not in the mood. I can't think of prancing around with you at a time like this. I've got more important things on my mind than you and your Saturday night escapades."

11. "When I asked her to go out with me, she actually laughed in my face. It was horrible. I felt like two cents."

12. Frank is a student in Mr. Davison's geometry class. He is having difficulties with applying the distributive property to proof construction so he asks Mr. Davison to explain it. Here is their discussion:

Frank: "Mr. Davison, could you go over why distribution can be used to solve the proof on page 200?"

Mr. Davison: [Roles his eyes] "That's a really dumb question. We've done about a thousand other problems like that. Anybody have any other questions?"

[No one asks any further questions]

*13. Tanya and Craig are both in their late thirties and have been married for ten years. Last night they had a quarrel. Tanya accused Craig of working too much and not spending enough time with her. She said that he was a "poor excuse for a husband" and that she was "sick and tired" of living with such a "boring" man.

Here is their discussion the next morning:

Tanya: "I just want to apologize for the things I said to you last night. Are you still upset with me?"

Craig: "Well, kind of. You really hurt me when you said those things to me. I couldn't sleep the entire night because of you. I hope you will not upset me like that again. I really should remain calm. I'm no kid anymore, you know. One of these days you're going to give me a heart attack!"

14. Carlton is a 40-year-old hardware salesman in a large department store. He has held the same job for six years and has not gotten a raise since he was employed. Last year, he interviewed for a sales position at a local hardware store, but he did not get the job. Recently, however, his brother-in-law, Salvador, offered to go partners with him in their own hardware store. Here is his response to Salvador's offer:

Carlton: "I'm sorry Sal, but, as much as I would like to do it, I just can't take you up on it. It's just too much for me. It's easier for me just to stay where I'm at. I just couldn't take the hassles of running a business."

15. Mr. Aimes, a community college instructor, is presently awaiting a tenure decision. Having had some problems with the college administration, he has doubts about whether he will be granted tenure. Here is a discussion he has with his wife:

Mr. Aimes: "What if I don't get it? Then what do I do?"

Mrs. Aimes: "Oh, I'm sure you'll find another job and, anyway, what's the use of worry now when you still might get it."

Mr. Aimes: "I don't think you even understand what I'm going through. This is my career we're talking about! How can you have the nerve to tell me not to worry about it? Don't you even care?"

16. Mother: "My kids are enough to make anyone crazy. When they don't listen to me, they get me so mad that I completely lose control."

*17. Dwayne has been in counseling for about two months and has been working on the avoidance of fallacies in his thinking. Here is what he tells his counselor in one of his sessions:

Dwayne: "This week I committed the fallacy of demanding perfection."

Counselor: "Tell me about it."

Dwayne: "My wife was supposed to meet me for lunch at twelve thirty and didn't show up until one. I got very angry at her. I told myself that people should never make other people wait like that, especially people who are supposed to care about you."

Counselor: "Do you think that you were making an unreasonable demand?"

Dwayne: "I obviously was. Now I feel like such a fool. At this point in counseling, I shouldn't be making such mistakes. Why did I let myself get so angry!"

18. Gregory is a 35-year-old novelist. He has been married for 14 years and has two children. In the past, he has published several books, one of which has become a best seller. However, he is presently experiencing "writer's block" and is unable to write anything publishable. Here is what he tells his agent:

Gregory: "When I write, I feel like I have a purpose in life; I feel like I'm worth something. But now I feel dead inside, like my life has lost its meaning. If I can't write anymore, I might as well be dead!"

19. "I'm a perfectionist. If I do anything then it's got to be perfect."

20. Deborah is a 15-year-old high school student. Her father had

sexually molested her from the ages 6 through 13. She is presently promiscuous and has a drug habit. Here is a discussion between her and a school counselor:

Deborah: "My father is the best father you could ever have."
Counselor: "Why do you think so?"
Deborah: "He buys me presents all the time. He's so good to me."
Counselor: "Has he ever done anything to hurt you?"
Deborah: "Oh no! He'd never do anything like that."

21. Music Student: "If I mess up in my solo performance tonight, then everyone will laugh at me, my reputation will be ruined and I'll be tortured for the rest of my life; I just couldn't stand it."

22. Mr. Shiner, a junior high school teacher, has problems with controlling his classes. Many of the students talk to each other while he tries to teach. Some of the students who want to learn are displeased by Mr. Shiner's inability to control the class. On one occasion, Mr. Shiner stops his lesson and says the following to the class:

Mr. Shiner: "OK, now I've had it with you animals. Don't you have any sense at all? Why do you come to school anyway? What purpose do you serve anyway in life, besides disturbing others?"
[Many of the students who had been talking laugh.]

23. "I refuse to eat in a restaurant by myself. I just couldn't stand sitting there, knowing that people were looking at me and thinking that I was some sort of oddball."

24. Here is an encounter between two men who are in love with the same woman:

John: "Why don't you crawl back into the gutter where you came from, you worthless scum!"

Bill: "Why you no good, dirty, rotten . . ." [Punches John]

25. Dr. Cynthia L. George, professor of mathematics, prides herself on her excellence in teaching math. However, on one occasion, she made an obvious mistake on the board, which was then corrected by a student in the class. Here is one thing she tells herself about this incident:

"I don't know how I could have made such a stupid mistake. I'm a full professor of mathematics; I shouldn't be making mistakes like this. Imagine being corrected in front of the whole class by a student. How terribly humiliating!

26. Son: "I was thinking about asking Mary Pine to the senior prom."

Mother: "Really? Does she still have those crooked teeth or did her parents finally invest in some braces? God only knows she needed them. She was always such a homely little thing."

*27. Alice and Pauline are friends and roommates at college. This evening Alice is planning to study for tomorrow morning's history exam when Pauline asks Alice to go with her to a party that same evening. Here is their discussion:

Pauline: "Hey Alice, let's both go to Lora's party. There's going to be some real cute guys there. Lora told me that Sam Larson is going to be there too. He has the nicest butt!"

Alice: "I don't think so. I have this history exam to study for and I really don't know the stuff."

Pauline: "Alice, you are such a bore! Well, go ahead and study for your dumb test. I guess you can't help it if you're a bookworm. Have fun with your history book."

Alice: "OK, I'll go with you. Can we be back by twelve?"

Pauline: "Sure, don't be such a worrywart. Let's get ready."

28. Judy is a cashier in a local bookstore store. Yesterday at work

she gave a customer back too much change, a ten instead of a five, and did not realize it until the customer had driven off. Here is what she tells herself as she is lying in her bed, still awake, at 3:00 A.M. in the morning:

Judy: "I know I really messed up big time. I can't stop thinking about what I did and how I should have known better. How could I have done such a thing."

FURTHER PRACTICAL EXERCISES FOR CHAPTER SIX

1. Dividing in smaller groups, let each group create and take turns *role playing* (before the whole class) at least one practice scenario containing at least one **Fallacy of Making Value Judgments**. The class may then critique the scenario, identifying the specific fallacy or fallacies committed and discussing how the characters in the scenario could have more effectively dealt with their situation. (*A Further Option*: After the class finishes its critique, the group may then "replay" the scenario, however this time omitting the previous fallacy or fallacies, dealing more effectively with the original situation.)

2. Collect at least 4 examples of at least 4 *different Fallacies* **of Making Value Judgments**. You may find these in newspaper editorials, magazine articles, advertise- ments, TV or radio commercials, political addresses, books, essays, etc. You may also use personal experiences you have had such as what you or somebody else said under a given set of circumstances. For each fallacy you find, write down the name of the fallacy and briefly describe why you believe it to be that fallacy. If you have *BELIEF-SCAN* you may use the program to support or establish your findings. Include in your project package copies of any supporting newspaper clips, advertisements, etc. that you have used (where available). Also attach any supporting *BELIEF-SCAN*

reports. Be prepared to present and discuss your findings in class.

3. Select an essay or editorial (from any source you want) written on a controversial subject, and analyze it for **Fallacies of Making Value Judgments**. Write a brief evaluation of it's strengths and weaknesses in the light of your findings. If you have *BELIEF-SCAN,* you may use it to check for these fallacies.

PART 2

KEEPING YOUR WITS ABOUT YOU:

RATIONAL THINKING

AND

EMOTION

CHAPTER SEVEN
ASSERTING YOURSELF

AVOIDING FALLACIES AS A GOAL OF ASSERTIVENESS TRAINING

Asserting yourself means exercising ***rational*** *control over your life.* It does not mean acting aggressively or in hostile ways toward others (for example, yelling at others or otherwise acting rudely) since actions such as these are likely to defeat your own purposes.

By working on identifying and avoiding faulty thinking, you are also working on your assertiveness skills. That is, when you stop committing a daily diet of fallacies, you will stop making so many irrational excuses for not *doing* those things that are likely to bring you greater happiness. However, even then you may still need to push yourself in order to overcome the force of inertia. So what else can you do?

SEVEN WAYS TO OVERCOME INERTIA

1. Put your actions where your mouth is

OK, suppose, for example, that you are now aware of how you have irrationally told yourself how your past must determine your future and that because you were unsuccessful with relations with the opposite sex in the past, you must be that way in the future. Now, what are you going to *do* about it? Don't just tell yourself how irrational you have been. The force of inertia is quite strong and you may *say* this to yourself without making any changes. So put your actions where your mouth is. For example, you might ask out that attractive person. "But she/he might turn me down," you say. Well, what if she/he did? Would that really be so devastating? You could always ask someone else out, or try again to get this person to go out with you; so *take some risks*, stop your slide down the slippery slope, and stop your awfulizing. The important thing here is that you work on adjusting your actions to fit your new beliefs instead of hanging on to the old patterns of behavior that your irrational beliefs have helped to sustain.

True, these irrational patterns of behavior (for instance, shying away from relations with the opposite sex) may have built up a momentum of their own over the years and have become habits. But you *can* change them if you push yourself hard enough to do something about it. Once you have decided to give up the fallacies that help to sustain your old patterns of behavior, there is really nothing to stop you. So don't fall back on the irrational beliefs; put yourself in gear, and act accordingly.

2. Be especially on the lookout for damning thoughts about yourself

What are you telling yourself when you withdraw from doing something you really want to do, for example, from applying for that job you always hoped would come along or from asking that very attractive person out on a date? Perhaps you are telling yourself that the job really isn't what you're looking for anyway or that the attractive person is probably "stuck up" anyway. But are these your real reasons for your inertia or are they merely rationalizations, that is, reasons you are thinking up in order to conceal, from yourself, more threatening reasons you think exist for not doing these things? But what are these other reasons that you could be trying to conceal? Often times they are ones that are based upon fallacies. For example, you may think that you might fail to get the job, or that the attractive person will reject you, and that if this or that happened, you would prove to be a failure. So instead of risking rejection, you sit back and you never make your move.

As the above examples illustrate, the fallacy of Damn-ation, especially as applied to yourself, can be, and often is, at the roots of inertia. But, since it may sometimes be buried beneath your rationalizations, you may have to look beneath the surface for it. Ask yourself if the reasons you are telling yourself (and perhaps other people) are honest ones or is your lack of initiative primarily due to your fear of the damaging effects you think failing would have on your self-concept.

3. Don't lock yourself into inertia for fear of how you think others might view you if your efforts were to fail

Sometimes inertia may be due to fear of how you think you would look to others if you were to fail. For example, "How would I look if I asked him/her to dance and, right in front of everyone, he/she said no."

But would anyone *really* think the worst of you if you asked him/her to dance and he/she refused your offer? What evidence do you have for thinking so? Without evidence to support your belief, you are committing the fallacy of Questionable Claim.

Suppose, however, that you do have good reason for thinking that at least some people might look upon you as some sort of flop with the opposite sex. Then what? Well, you say, "I mustn't have anyone think things like that about me." Well, why not? Is there some sort of natural law, stating that everybody *must* accept, approve, respect, love or admire you? Of course not. You are then just Demanding Perfection (In This Imperfect World).

Maybe your fear of how you would look if your efforts failed stems from your fear that others might find out what you really already know, namely that you are nothing but a big dud, which you have merely done a good job of concealing up until now. But, if this is your reasoning, then you are obviously committing the fallacy of Damnation.

So watch out for the "But what would I look like if I failed" excuse. A careful inspection of your reasons for making this excuse is likely to turn up some underlying fallacy.

4. Please yourself

Do you find it difficult to do anything for yourself unless it also pleases other people in some way? Does it seem to you as if there is usually little or no time to accomplish anything you want to do because you are too busy meeting other people's demands? If this description comes close to the way in which you would describe things, then it is time to take a look at some of the irrational things you might be telling yourself in order to put yourself in this bind.

Why in the first place might you routinely choose to do what others would prefer instead of doing what you would prefer? One possible motive is fear that you would otherwise lose the approval of others, which you irrationally think would reflect badly on your personal worth. But, as you have seen in Chapter 6, it is a fallacy (a form of Damnation) to make your personal worth dependent upon other people's approval. So, if this is what you are telling yourself, give it up!

Here are some other possible things you might be telling yourself: "I must please others first," "If I put myself first, then I am being selfish," "Whenever I do anything for myself, I am being selfish." Why would you believe these things, however? These things are not self-evident and if you have assumed that they are, then you have committed the Self-Evidence Fallacy. There is simply no universal, self-evident, law, moral or otherwise, that says that you must always please others first. If it were a universal law that people always sacrificed their own happiness for the sake of the other person's happiness, then nobody would end up being happy. For example, parents cannot rationally assume that they must always sacrifice their happiness for the sake of their children's happiness. If this were universally accepted, then the children of unhappy parents would themselves grow up to be unhappy parents. What a wasteful sacrifice of human happiness that would surely be!

It is also irrational to think that whenever you do anything for yourself you are being selfish. To have acted selfishly means that you have violated the rights of others. Since doing something for yourself need not violate the rights of others, such acts need not be selfish. There is an important difference between being selfish and being self-interested. You can be interested in yourself without being selfish. For example, you are not being selfish when you do something for yourself like taking a class or working toward a degree or spending your day off doing something you enjoy. Remember that you, just like others, have a right to your personal happiness and you are not being selfish just by virtue of exercising that right.

So, don't let these irrational ideas stop you from doing what pleases you, even if they don't happen to satisfy the agendas of others. For example, if you like to read, then you might go out and buy a book that interests you. If you enjoy sailing, then you might go sailing. Finding the time to do more of the things you want to do is likely to be a lot less difficult once you get rid of the irrational idea that you must sacrifice your own happiness for the happiness of others.

5. Learn how to say no

If you are inclined to say yes to most things you are asked to do, even if you have good reasons for not wanting to do them, then you are likely to find that much of your life is taken up by meeting the demands of others and that very little time is spent pleasing yourself. So one thing you can do is to learn how to say no.

There is, of course, a difference between being hostile towards others who make unfavorable requests of you and asserting yourself. You can politely refuse to do something by explaining to the person making the request your reasons for not choosing to do it. Moreover, where you have a right to refuse a request, there is no point to apologizing; by refusing in such cases, you have not done anything wrong.

Since saying no may be a new behavior for you, you may need to push yourself to do it, at least at first. With practice, however, it is likely to become easier for you. It is a good idea for you to work on this new behavior by assigning yourself occasions where you will practice it. For example, you might be prepared for the next time your coworker tries to exploit you by asking you to bring him/her a cup of coffee. When he/she asks, you can then say no, explaining that this is not any part of your job description and that he/she could always go get the coffee himself/herself.

6. Avoid making your decisions by indecision

A common source of inertia is what is sometimes called "decision by indecision." This happens when you put off making a decision on some matter until the matter is finally settled with time and changing circumstances. For example, suppose you are trying to decide whether or not to accept a new job. If you go too long without deciding, then someone else may be hired for the job. You will have then made your decision by indecision.

A major problem with making your decisions in this way is that you thereby lose the opportunity to exert rational control over your life. Instead of making decisions concerning your future well-being (and that of others) on rational grounds, you thereby allow things to happen without your influence and direction. While occasionally things do happen for the best, it is unlikely that, by leaving things to the random unfolding of events, you will "luck out." More likely it is that you will miss important opportunities for affecting constructive changes in your own life as well as those of related others.

How, then, can you avoid decision by indecision? One clear

way is to watch your fallacies. For example, if you magnify risks and awfulize about them, you may end up making your decision by indecision. Suppose you are undecided about the purchase of a new home. You tell yourself how nice the home is and how much you would like to own it. On the other hand, you tell yourself that it is less risky just to stay in your old home and how awful it would be if you lost your job and couldn't make payments on the new home. So, you might keep yourself in limbo over whether to purchase the new home. Indeed, you may end up spending the rest of your life in your old home without ever having decided to, unless of course you take rational control of your life and make a decision.

7. Avoid dilemma thinking

A related way in which you can get caught up in decision by indecision is to think of decision making in terms of resolving hopeless dilemmas. A dilemma is a situation in which you can do one thing or another; but in either case, your goose is cooked. It is a case of "damned if you do, and damned if you don't." For example, suppose you and your mate are undecided about whether or not to have children. Since the biological clock is ticking at an advanced stage, you feel somewhat pressured to make the decision. You reason, "Either we have kids now or we'll never have any. But if we have kids, then we'll never be able to travel and we won't have the freedom to do whatever we want whenever we want. On the other hand, we'll probably regret it if we never have any kids. When we're old we won't even have any grandchildren. What a real dilemma!"

But would this, indeed, be a *real* dilemma? Not really. First, you would be committing a "false either or" since there are other options besides having children now and never having any at all. Obviously, you could always try to adopt. Moreover, you may irrationally be assuming that you are both so fertile that conception will occur right away. And even if you really are that fertile, there is, after all, a gestational period, which means that you could not have children immediately anyway.

Look too, for example, at the claims that you will *never* be able to travel and you will not have the freedom to do *whatever* you want *whenever* you want. A moment's reflection shows that these claims commit fallacies. The first claim commits the Slippery Slope fallacy by exaggerating the effects of children on the potential for travel. The second claim commits the Straw Man fallacy by assuming that, without children, you have the freedom to do *whatever* you want *whenever* you want. Even childless couples have some restrictions on their freedom.

A danger of thinking in terms of dilemmas such as the above one is that it can lead you to make your decision by indecision. By repeatedly telling yourself that you are damned no matter what you do, you may never reach a verdict. For example, you and your mate may, through indecision, never have any children.

Many times, as in the above example, dilemma thinking contains fallacies. Therefore, by avoiding fallacies, you will be less likely to engage in dilemma thinking. And, since dilemma thinking can, as you have seen, lead to decision by indecision, it is a good idea to avoid such thinking. How- ever, if you find yourself "damning if you do and damning if you don't" be careful. Take a critical look at your dilemma, for it might not be a real one. While not all dilemma thinking contains fallacies, much of it does. So, remember to probe for fallacies, and watch out for the perils of decision by indecision.

PRACTICE SCENARIOS FOR CHAPTER SEVEN

IN EACH OF THE BELOW CASES, SOMEONE IS HAVING DIFFICULTY IN ASSERTING HIMSELF/ HERSELF. WHAT DO YOU THINK EACH OF THESE PEOPLE IS DOING WRONG AND WHAT MEASURES WOULD YOU SUGGEST THAT EACH TAKE TO CORRECT HIS/HER DIFFICULTY?

1. Joe had been attracted to June for the past 6 months, and had been trying to decide whether or not he wanted to get involved in another relationship. Recently, he has heard that June has begun to date Jim, Joe's best friend. Joe now believes that this probably happened for the best because he probably wasn't ready for a new relationship anyway.

*2. Pat, a computer science instructor at a local college, is married with two children. He has been hoping to finish writing his doctoral thesis since this would mean an increase in salary. However, every time he begins to have a break in his work schedule, where he might be able to do some writing, one of his colleagues asks him to take on another assignment. Pat is eager to please and to be well thought of by his colleagues so he "gracefully" accepts. As a result, he is on more college committees than any other faculty member. In fact, the other faculty members have come to associate his name with the slogan, "Need help with that? Just call Pat."

3. Ronna and her husband, Burt, are discussing an en- counter that Ronna recently had with her employer, Mr. L.

Ronna: "Burt, I really blew it today at work."

Burt: "What happened?"

Ronna: "Mr. L. asked me to type several documents pronto because he needed them for the New Mexico meeting tomorrow. I just about had all of them done when I spilled my coffee all over two of them."

Burt: "So, did you type them over?"

Ronna: "Yeah, but I was a half hour late in getting them to Mr. L. He just looked at me kind of funny, so I apologized for being late and asked Mr. L. if I could work overtime this week to make it up to him."

Burt: "Work overtime, Ronna?"

Ronna: "Don't worry Burt. Mr. L. said it wasn't necessary, but I just can't stop thinking about what a dumb fool I am. Imagine, I can't even type a few documents without messing up. I know Mr. L. will never consider me for a promotion now. I'm not even going to ask him about it."

4. Throughout the five years that he has been employed by his company, Mack has made it a point to comply with every request his boss makes. This includes staying after hours to help complete work that Mack's coworkers have not completed. Recently, Mack was asked to work late on the evening of his wife Lidia's birthday. Although Mack knew that Lidia was expecting to go out to dinner, Mack was certain that his boss would view this as a frivolous excuse. When Mack returned from work at 10:00 P.M., Lidia was furious.

5. Pearl and Pete are experiencing difficulty in deciding whether or not they want to purchase their own home. They both agree that they are presently living in cramped quarters but worry that they will be stuck with a house should they decide to move out of state in a few years. Both Pearl and Pete have now become overwhelmed even discussing the matter, but they believe they must decide now. As Pete put it, "either we make up our minds to buy a house and be stuck living here forever, or we remain living in this cramped two room apartment and stop complaining."

6. It has been two years since Mary graduated from college. She has repeatedly told both herself as well as family members that she will use the skills and knowledge she has acquired to start a successful accounting firm. Mary oftentimes spends hours formulating plans for this business venture, but she says that her present job as a check-out clerk at the local grocery store leaves her too tired to undertake her business now. However, she declares, "Someday soon I will get up the energy to go ahead with my plans."

*7. Sonia was married once and divorced. Although her husband had an affair and left her, Sonia has told herself that it was probably her fault since she did not "please" him enough and that she really didn't "deserve" a man like her husband anyway. Now, every time another man tries to get close to her, she breaks it off, saying, "It wouldn't have worked anyway."

8. Irene is a 50-year-old widow. Her husband died almost three years ago. George, a widower whom Irene had known since high school, had, on several occasions over the past couple of years asked Irene out. Irene, however, refused his offers saying that she was really not ready for another man in her life. She told him that she would let him know when she was ready. Six months went by, however, and George did not call. In that time, Irene had been thinking how much she really liked George, and how she just might take him up on his next offer. Thus, when he finally called yesterday to tell Irene that he and Iris were to be married, Irene had a difficult time pretending she was happy for them both.

9. Perry has been employed at the same job for nine years. In that time he has worked long hours, has received relatively low wages, few benefits, no raises, and, what is more, he has found the work extremely boring. While, on many occasions over the years, he has told his wife that he is "fed up" with his job and that he "ought to have his head examined" to stay all those years at a job like his, he has not attempted to find suitable employment elsewhere.

*10. Kathy is married, a homemaker, mother, and a part-time sales clerk. She has two children and her 94-year-old mother-in-law, an invalid, lives with her in a small three- bedroom home. While Kathy used to enjoy playing tennis, swimming, and dancing, she has not engaged in any of these activities for some time now. She says she has little time these days for such "frivolous things." After all, she says, "I just can't go off and play tennis when others are depending on me. How could I be so selfish? I could never respect myself, if I did anything like that."

11. Benny, a routes man for the U.S. Postal Service, enjoys cooking and often thinks about opening his own restaurant. He has also spent many of his days off checking out "prime locations" and making "preliminary inquiries." However, whenever he gets close to going through with his plans, Benny changes his mind saying, "Its a dumb idea anyway. I'd probably end up losing my shirt." Nevertheless, Benny still keeps looking for prime locations.

12. Olive and Frank, her husband of ten years, have three children. They both come from close-knit families and live near their parents and siblings. Olive's daily schedule consists of transporting the children to and from school, to their extra- curricular activities, grocery shopping for her two sisters who are employed full-time, and watching two of her nieces on Tuesday and Friday mornings. Olive spends each weekend visiting her parents and her parents-in-law, as she believes this will make them happy. Olive's free time is spent doing cooking and housework. She prides herself on having a clean house and on being a fine cook. However, she has recently begun to cry for no apparent reason and has begun to argue with both of her sisters and with Frank. Frank has suggested that Olive may be coming down with a virus.

*13. Elias and Geraldine have been married for four years. They have both agreed not to have children until Elias has finished his college education and can get a "decent" job in his field, which is teaching. Now that Elias has graduated from college the two have talked about having a baby. However, they presently rent a small apartment and would also like very much to own their own home. For several weeks they have debated the pros and cons of what to do first: purchase a new home or try to have a baby. According to Elias, either they can do one or the other thing first. But if they choose to have a child first, then it will be years before they will be able to afford to buy their own home. On the other hand, if they purchase a new home, then they won't be able to afford to support a child.

14. Penny, age 32, had been a high school honor student but had married early and had never gone to college. She has secretly wanted to return to school for the last four years but is afraid that her husband and mother-in-law may think she is shirking her responsibilities as a mother if she does return. She is especially fearful that she will disgrace herself if she does go back against their objections and does poorly. Penny continues to daydream about going back to school and becoming a teacher, but she has told herself that this dream will never come true.

CHAPTER EIGHT
REDUCING YOUR STRESS

STRESS AS INTENSE EMOTION

Stress can be defined, for practical purposes, as any *intense emotion*. While some intense emotions may be beneficial and appropriate (for example, intense fear of a wild boar heading your way can move you to take cover or defend yourself), many intense emotions that people experience are both harmful and inappropriate. They are harmful because they can interfere with effective thinking. For example, intense anxiety can impair your ability to concentrate, remember, and learn. They are also harmful because they can, over time, do serious physical damage. For example, ulcers, colitis, heart attacks, high blood pressure, and a number of other physical maladies have been linked to stress.

Stress is *inappropriate* when your emotional response to an environmental stimulus is self-defeating and interferes with your ability to deal effectively with your environment (work, study, interpersonal relations, etc.). Since stress amounts to intense emotions, stress management may be conceived of as the control of inappropriate, intense emo- tions.

All intense emotional responses to environmental stimuli typically involve three principle changes: (1) changes in your thinking--your factual claims as well as your value judgments, (2) physiological (bodily) changes, and (3) changes in your behavior. Thus, for example, if another motorist cuts you off as you are driving down the road, and your emotional response is one of intense anger then (1) you are likely to think certain things ("That no good SOB! He could have killed somebody!"); (2) your adrenalin will start pumping and your heart and respiratory functions will accelerate; and (3) your behavior will change (for example, you might yell something out the window or make a certain hand gesture.)

These three elements of emotions are *mutually interactive.* This means that each can influence the other, and vice versa. For example, changes in your thinking can cause bodily changes; and bodily changes can cause changes in your thinking; as when you think a "bad" thought and your heart starts thumping; or, when a bodily change due, say, to medication or lack of sleep, affects your ability to concentrate.

Corresponding to the above three interactive elements of emotions are three methods of stress management. First, you can manage your stress by working on your thinking, second, by drug therapy, and third, by behavior therapy.

The third method involves avoiding stress by changing your behavior, that is, by learning new, more constructive behavioral responses to life situations. For example, you can change your study habits so that you no longer put yourself in the situation of having to cram; or, for example, you can confront the fear of rejection by going ahead and doing the dreaded thing, say, asking someone out on a date. While this method is a useful and inexpensive *adjunct* to the first method--that of working on your thinking--it is no substitute for it. For example, if you are unable to study in advance due to circumstances beyond your control, you will still need to deal with the situation at hand; or if the person whom you ask out turns you down, you will still need to deal with the rejection.

While the causes of stress can and do include biological factors for which the second of the above methods (drug therapy under the care of a *competent* physician) may sometimes be indicated, inappropriate stress is often largely a product of irrational thinking about life situations. In *these* cases, you can deal constructively, inexpensively, and safely with stress by giving up some of the *fallacies* in your thinking and, so, by thinking more rationally. It is this first method of stress management as *the reduction of faulty thinking* that will be the primary focus of this chapter (though not to the exclusion of the second, behavioral method).

For instance, if you keep telling yourself that, either you do well at your job interview tomorrow, or you might as well be dead, then you are likely to be stressed out. Dealing effectively with stress in this case would mean stopping your false either or-ing and drawing more rational infer- ences about the job interview--for instance, "it wouldn't be the end of the world for me even if I did mess up."

As you become more adept at getting rid of fallacies in your thinking, you are likely to notice how situations that previously got you uptight are not quite so stressful as they used to be. You are likely to *feel* better and, as a result, do better in other respects, for instance, on the job or in relating to others.

But are there any particular kinds of fallacy I should especially look out for if I find myself experiencing intense emotional stress? While *all* the fallacies you have examined can and do contribute to your feeling stressed, there are certain fallacies that appear to play a central role in causing specific *intense* emotions. Many of these "stress fallacies" belong to the two groups of fallacies known as Fallacies of Fact Gathering and Fallacies of Value Judgments, which you have tackled in Chapters 2 and 6, respectively. Let's look briefly at some of these fallacies as they relate to specific kinds of emotional stress.

ANXIETY, ANGER, DEPRESSION AND GUILT: WHAT FALLACIES TO LOOK FOR IN EACH

1. If you are experiencing intense anxiety, look for:

The Slippery Slope-Awfulizing-I-Can't-Stand-It-Itis Syndrome

Intense anxiety frequently arises as a result of negative thinking about imagined future events in your life. If you think that something *awful* will (or might) happen to you, then you are likely to be extremely anxious about it. For example, first you tell yourself that your boss is going to fire you tomorrow because you lost that account today. Then you tell yourself that you are finished, washed up, in your field. Nobody is ever going to hire you now. Then you tell yourself that you'll have to get a menial job somewhere, making a fraction of the salary you are now making. Then you tell yourself how you'll have to sell your house and move into some roach infested apartment, and give up the standard of living you have become accustomed to. Then you start telling yourself how awful this would be and how you just couldn't stand it. Are you now anxious about going to work tomorrow? It would be hard not to be after selling yourself this can of worms!

So watch out for the Slippery Slope-Awfulizing-Can't Stand-It-Itis Syndrome. It is a real anxiety producer. A useful way of checking for it is to ask yourself a series of "what then?" questions. (See Chapter 1.) For example, you can tell yourself, "OK, so I lost the account. Then what? Well the boss will fire me? Then what? I'll be finished in my field. Then what? . . ." Notice how, in this manner, you can lay out your own reasoning for your inspection. This will give you an opportunity to rationally challenge that reasoning, and to discover the fallacies in it. For example, "What evidence do I really have that the boss is going to fire me? And even if he does, how do I know that I couldn't find another suitable job in my field? And even if I couldn't would it really be so awful? I might even be able to go back to school and prepare for a more rewarding career in something else. Alas, am I not just falling into the old SlipperySlope-Awfulizing-I-Can't-Stand-It-Itis Syndrome?"

Of course, it would be best to think rationally in the first place about your circumstances. But, even if you don't, you can still catch yourself, by asking yourself the "What then?" question, and by rationally inspecting your answers.

2. If you are experiencing intense anger at somebody else, look for:

Damnation, Demanding Perfection, Personal Attack, Two Wrongs Make A Right, Appeal To Force

When you are very angry at somebody, chances are that you are saying some very negative things about them as persons. For example, your spouse has just suggested that you go on a diet and and here is your reaction: "Why you insulting SOB. You have no right to insult me like that. Especially you, you fat ugly horse! If you ever criticize me like that again then I'm out of your life forever." But look for a moment at this twisted logic. You are saying: your spouse is an SOB (Damnation) because he/she said something negative about you, and he/she must never do that to you (Demanding Perfection). Moreover, since your spouse is a "fat ugly horse" (Damnation again), he/she is not qualified to comment on your weight (Personal Attack); and its alright for you to let him/her have it with both barrels and put him/her in his/her place because he/she did something to you that you perceive as being wrong (Two Wrongs Make A Right); and just to make it look like you're in control, you threaten to leave her if she doesn't watch what she says to you (Appeal To Force). While different fallacies may also come into play in situations where you are very angry at another person, at least some or all of these fallacies are likely to be present, so you might do well to be on the lookout for them in such situations.

3. If you are feeling depressed over some event or circumstance in your life, look for:

Self-Damnation, Awfulizing, I-Can't-Stand-It-Itis, Magnifying Risks, Straw Man, Slippery Slope, Had Things Been Different, Neglecting the Past, False "Either Or", "Terrific" Fallacy

Since depression may require medication, it may be necessary to be under a physician's care. Nevertheless, when you are depressed, you are likely to commit at least some of the fallacies listed above (but do not assume that the range of fallacies you might commit while depressed is restricted to these alone).

Depression is characterized by a bleak outlook, which, in some cases, may take the form of Self-Damnation. For example, after losing your job, you might tell yourself that you are a big failure and that you can never do anything right. In this case, working through your depression will involve giving up the irrational conviction that you are an equation for failure.

The bleak outlook of depression might also be directed upon life circumstances in the forms of Awfulizing and I-Can't-Stand-It-Itis. For example, after the death of a loved one, you might tell yourself how horrible it is that this has happened and that you just can't stand to live anymore, that you might as well be dead too. In this case, working through your depression will involve coming to appreciate the unfortunate character of your loss while also coming to see that it is still not the end of the world for you, and that you can, after all, stand it.

When you are feeling depressed, look also for Fact Gathering fallacies such as Magnifying Risks, Straw Man, Slippery Slope, "Had Things Been Different," and Neglect- ing the Past. For example, you may tell yourself that *nobody* really gives a *damn* about you (Straw Man); that you will *probably* die anyway before you reach forty (Magnifying Risks); that, because your spouse left you, you are going to end up alone for the rest of your life (Slippery Slope); and that had you tried harder to please your spouse, your marriage would not have failed ("Had Things Been Different"). When you are depressed, you may also commit the fallacy of Neglecting The Past by failing to recall and appreciate the way problems similar to the one you now face had satisfactory resolutions. For example, you may fail to take account of how the last time you lost your job, you were still able to get another job despite bad economic times. Keep in mind that the stress of depression can dull your mental abilities, including your memory, so you will need to make a special effort to focus upon the facts in order to avoid these various Fact Gathering fallacies.

In a depression, you may also do some "damned if you do and

damned if you don't" thinking (dilemma thinking), which was discussed in Chapter 7. If so, in addition to the above fallacies of Fact Gathering, look for the False "Either Or." For example, you may tell yourself that you can either stay unhappily married or get a divorce and be alone for the rest of your life, so that you are damned no matter what you do. There are, however, other possible options in between these such as working to improve your marriage, going to a marriage counselor, or, if these things don't work, trying to meet somebody else. So look for other options beside the negative ones upon which you are likely to focus when you are depressed.

When you are depressed over a personal loss, you may also commit the "Terrific" Fallacy by telling yourself how wonderful what you had was, and so how awful it is that you have lost something that wonderful. In your depression, you may be inclined to see only the good features of what you have lost. For example, even though you may have had many pleasant times in your marriage, there were also the unhappy times; nor was your spouse a saint!

4. If you are feeling intense or debilitating guilt, look for:

Self-Damnation, I-Can't-Stand-It-Itis,
Awfulizing, Demanding Perfection,
The "Thou Shall Upset Yourself" Fallacy

Guilt is a *moral* emotion. That is, whenever you ex-perience guilt, you are telling yourself that your action, or inaction, was morally wrong, or that you have violated some moral rule. (Note that you can feel guilt if you *think* that you have done something morally wrong even if you did not.) Now, if you stick to rating, as morally wrong, what you have done rather than your entire self for having done it, you may be motivated to change your behavior in the future. On the other hand, if you rate your *self* as being totally bad, evil or wicked, then your attention will not be on the behavior in question, but instead on your self. But how can a totally bad self ever hope to do anything right!? In other words, by damning yourself, you will self- defeatingly succeed in thwarting your own powers of con- structive change.

Notice too how your self-damnation may be supported by other fallacies such as Awfulizing and I-Can't-Stand- It-Itis. For example: "It was horrible of me to have spoken to my mother like that, especially since she's always been so good to me. I just can't stand to think that I did such a thing. What a no good heel I turned out to be!" In this case, you are likely to experience intense and debilitating guilt. On the other hand, you can avoid the Awfulizing, Damnation and I-Can't-Stand-It-Itis. Instead you can tell yourself that, "I know I said some things to my mother that I shouldn't have said to her, but I can always apologize to her and try to learn from my mistake. After all, I'm not perfect and what I did was not horrible. And, in any case, the fact that I have done something wrong does not mean that I am no good." In this case, you are less likely to experience intense guilt and to act more rationally regarding your mother.

Another route to irrational guilt is to make unrealistic absolutistic moral demands upon yourself. For example, you might tell yourself that you must *never* break promises or harm others or tell lies. However, if you tell yourself such things then you are committing the fallacy of De- manding Perfection since these absolutistic demands cannot reasonably be met, at least in this imperfect world of yours. Sooner or later, no matter how hard you try to make good on them, they will come into conflict and you will have to give up one of them. For example, you may have to break a promise or tell a lie in order to avoid harming someone else. Moreover, it is plainly unreasonable to expect yourself, as a fallible human being, to never do anything morally wrong. Do you know of any human (not a god!) who has *never* done anything morally wrong?

You are less likely to experience intense guilt when you break a moral rule if you think of such a rule as admitting of exceptions. For example, it may be *usually* wrong to break promises, but that does not mean that there are no exceptions to this rule. Thus you would not be acting wrongly if you broke your promise to your child because keeping it would harm him or her. And if you understand that you have done nothing wrong under these special circumstances, then you are less likely to feel guilty (although you may feel some other emotion such as regret). On the other hand, if you accept the irrational philosophy that promises must not be broken under any circumstances, then you ar likely to experience intense guilt.

Watch out also for the "Thou Shall Upset Yourself" fallacy in the form of insisting upon a special duty to make yourself feel intense guilt for any of your misdeeds. By imposing such a duty upon yourself, you are likely to take one guilt trip after another thereby leaving yourself little opportunity for happiness.

Can you live an entirely guilt-free existence? Don't count on it. But by working on the fallacies that promote and sustain intense guilt, you may get rid of much of the guilty feelings that you presently suffer.

Is a guilt-free life ever a good thing to have even if it could be achieved? Guilt can sometimes motivate you toward taking responsibility for your actions and toward affecting positive change in your life. In such cases guilt may be a rational emotion. However, when it involves faulty thinking such as Self-Damnation, I-Can't-Stand-It- Itis, Awfulizing, Demanding Perfection, and the "Thou Shall Upset Yourself" fallacy, then the emotion of guilt is likely to be debilitating and self-defeating and thus one that you would do well to avoid.

SIX ADDITIONAL WAYS TO REDUCE STRESS

Granted that you can make yourself feel better by catching and rationally disputing the fallacies in your thinking, but is there anything else you can do to get rid of stress in your life?

Yes, there are other things that you can do, but don't expect to get rid of all your stress. First, since some stress can be useful, this is not desirable. Second it is not even humanly possible! So, with this in mind, what else can you do?

1. Do what you can about your problem

There is no use in letting yourself stew in your problem when there is some constructive way of dealing with it. For example, instead of repeatedly telling yourself that the exam next week is going to be a real killer, you can omit the awfulizing and begin to prepare for it. By letting yourself stew in it, without attacking the problem, you are likely to increase your anxiety over the exam, which would obviously make matters worse.

Of course, a constructive solution may not always be possible or you may have already done all you reasonably can do about your problem. So what else can you do to decrease your stress?

2. Distract yourself

If there is no other reasonable action that you can take to ease your problem, then it is self-defeating to continue to focus your attention on it. You might then better change the subject. Focus your attention on something else that is less problematic. For example, instead of sitting next to the phone waiting for that important call, you can read an interesting book, or get yourself involved in some activity that you enjoy; or, you can simply carry on with your daily routine. Recall that you have no special duty to preoccupy yourself with, and upset yourself over, difficult situations in your life. So it is irrational to feel guilty about diverting your attention from the problem.

"But how do I know that there really is nothing else I can do about my problem? Maybe I have overlooked some important point, which might come to me if I would only keep going over and over the problem in my mind. By distracting myself, I might fail to think of something im- portant!" Keep in mind that there is a difference between giving due consideration to a problem and irrationally beating it to death. (In this case, you would be Demanding Perfection by insisting that you not overlook anything, and you would also be committing the "Thou Shalt Upset Yourself" fallacy by commanding yourself to stew in your frustration.) Keep in mind also that by irrationally telling yourself that you *must* preoccupy yourself with your problem so that you don't overlook *anything*, you are likely to increase your stress. However, since your powers of thought are diminished under stress, this strategy is likely to be self-defeating. Insight is more likely to occur when you are in a more relaxed state of mind. By focusing your attention on less problematic things, you might reduce your level of stress and be more likely in the end to effectively handle your problem.

As was noted earlier in this chapter, changes in *behavior* can sometimes help to relieve stress. Thus, if you are sitting at home telling yourself how awful, horrible and terrible things are, you are likely to feel quite down. However, by *pushing yourself* to get out and *do* something else that you sometimes enjoy doing, *even if* you don't presently feel like doing it, you can feel better and do better in the long run.

3. Schedule yourself

One popular occasion for stress is the case of too many things to do and not enough time in which to do them. If this is how you perceive your life, then perhaps you need to schedule your time more efficiently. Make a schedule of the things you intend to do along with the times at which you intend to do them. By doing this, you will be less likely to let things go to the last minute wherein it becomes difficult or impossible to deal with them all. It is advisable that you get a calendar book for the current year, with ample space for recording and describing your appointments, projects, and other items you plan to do. Some people write things down in many different places, for example, on many pieces of scrap paper (which they leave in many different places). When they need the information, however, it becomes difficult for them to find it, and so they experience unnecessary frustration. Moreover, get into a habit of consulting your calendar of events. If you never look at it, then why keep it!

Give yourself enough time to accomplish your tasks and, wherever possible, leave some additional time for the unexpected. Remember that things don't always go according to plan, so don't demand perfection. If you get off schedule, then do your best at rescheduling.

4. Don't take on more than you can handle

Be careful about overbooking yourself. Before taking on something new, consider carefully whether it is something for which you have adequate time. If it is something that is important to you, but scheduling is tight, then you might try to accommodate it by either putting off or giving up something of lesser importance. In any case, taking on more than you can reasonably handle, within a given period of time, makes it more likely that you will experience stress. Therefore, know when, and how, to say "No."

5. Avoid rating yourself

As you have seen in Chapter 6, it is a fallacy to damn your entire being on the basis of your behavior. It is, you have seen, more rational to rate your behavior, leaving your self out of the rating. But isn't it OK to give myself a positive rating when I behave in a commendable way?

From the perspective of stress management, a plausible answer to this question is no. When your self-worth is made a function of your deeds, then you are likely to experience anxiety whenever you think that you might fail at what you are doing. Today you might do well and think highly of yourself (for example, "I'm a worthy person because I did well on my exam today"); but what if you do poorly the next time around? Then you will no longer be able to give yourself a positive rating. You will live from day to day with the fear of whether or not you will retain your positive self-status or fall to the depths of complete worthlessness. And is this stress really worth it?

An alternative to making your self-concept a function of your deeds is to give your self a positive rating regardless of what you do ("I'm still OK even if I fail"); and then restrict your ratings to your deeds. Even though you have failed (for example, you lost that beauty contest) you still remain a worthy person in your own right. But why is it reasonable to always accept myself?

For one thing, you are likely to live with less stress that way. However, if what you want is some deeper philosophical reason for giving yourself unwavering value (what philosophers sometimes call "intrinsic value"), it is not hard to find such a reason. If you believe in God, then you might see yourself as valuable because you are one of God's children. If you are an atheist or do not especially believe in God, then you can still see yourself as valuable. After all, you are no mere chunk of lifeless earth; you are a living conscious being who thinks and has feelings.

6. Accept yourself as an imperfect being

As long as you demand perfection of yourself, you are likely to be very frustrated, indeed; for this is not even a human possibility. If you tell yourself that you *must* never make any mistakes, then you will inevitably make your self-worth depend upon your deeds ("Either I do it right or I'm no damn good"); and as you have just seen, this is a royal route to stress. So for the sake of avoiding stress, it is better to *give yourself permission* to make mistakes.

One special case of demanding perfection that is worth emphasizing in connection with stress is one that arises out of trying to correct the fallacies in your thinking. If you tell yourself that you must never mess up, lose you head, act rashly, overlook evidence, make bad inferences, and so forth, then you are likely to encounter a good deal of stress as a result of trying to live logically. This, however, is itself illogical, since, as you have seen (in Chapter 6), it is illogical to demand that you be perfectly logical. So, be prepared to make mistakes, and do not damn yourself when you do.

PRACTICE SCENARIOS FOR CHAPTER EIGHT

IN EACH OF THE BELOW CASES A PERSON IS EXPERIENCING STRESS. IN WHAT WAYS HAS THIS PERSON CONTRIBUTED TO HIS/HER OWN STRESS, AND WHAT MEASURES DO YOU THINK THIS PERSON MIGHT TAKE IN ORDER TO REDUCE HIS/HER STRESS?

1. Jeff has recently accepted a job as a psychology instructor at a local college. He has just gotten out of graduate school and has had no prior teaching experience. Since he received notice about the job, Jeff has been unable to sleep at night. He says that he "stays up most of the night in bed thinking about the job and just can't seem to fall asleep."

2. Ethel lives fifteen miles from her place of work. Every morning she gets up at 7:10, gets ready for work, and is out of the house and in her car by 8:10. Since she is due at work by 8:30, she finds herself speeding to work. When traffic is normal she finds that she has enough time to get to work. However, on several occasions she was almost killed by attempting to pass slow-moving cars traveling in front of her on a narrow winding highway. When she must stop for a red light or a train, she says she feels "like exploding." By the time she gets to work she says she is a "complete basket case."

3. Harry and his wife Henrietta are frequently embroiled in heated

arguments. For example, last week Henrietta called Harry an "inconsiderate SOB" when he came home from work late one evening without phoning to say that he would be late. Harry responded by telling her that she is nothing but a "damn complaint artist," that she should "go to hell," and that unless she cuts it out, he is definitely going to divorce her this time. So far, however, the two have gone on for ten years of marriage in this way and both have described their relationship as a very stressful one. According to Harry, even when they are not fighting, he still feels "uptight" since he knows that sooner or later she is going to "start up with him." Henrietta, however, says the same thing about Harry.

4. Janie, a single parent, often yells and screams at her children for not obeying her, for example, for not cleaning up their rooms when they are told or for not going to bed on time. She tells them that they are no good and that they will never amount to anything. She also threatens to "ring their little ugly necks" if they don't start listening. She says that those "little monsters" make her so upset that she feels like she's "gonna have a stroke."

*5. Jerry, a construction worker, has recently fallen off of a roof while on the job. He is presently in a wheel chair. His physical therapist has told him that, if he is to walk again, he needs to do his exercises. However, Jerry has so far not cooperated. Instead, he repeatedly keeps telling himself how horrible it is that he can't walk and how he'll never be the same again no matter what he does.

6. Johnny has been asked to be president of a social organization in which he has been active. While he is not keen on the idea, he accepts it because he doesn't want to "let the other guys down." He holds a full-time job during the day and also a part-time job at night so that he can, as he puts it, "have enough money to live right." He is married and has three children and he is a Cub Scout leader. It has been six years since he has gone anywhere on his

vacations, which he has spent taking care of things that "have to be done." Recently he has been complaining of headaches and dizziness. He also complains that, "there are not enough hours in a day to get done what needs to be done."

7. Dan is a psychiatric nurse at a local hospital. He tells himself that he is a professional and that he must always do what's best for his patients. He says that he "can't afford to make any mistakes." The last time one of Dan's patients committed suicide, he went into a lengthy depression. Dan often complains that it is difficult for him to relax even when he is not at work.

8. Ken is a guitarist and lead vocalist in a rock group. His band has recorded two records so far. The first one was a big success. After the record reached number 2 on the charts he told his wife that he really felt good about himself for the first time in his career: "People are finally starting to think something of me. I love it!" However, the second record did not do well and Ken went into a deep depression, declaring "I guess I really never was much good anyway. I'm nothing but a has been who never really was."

9. Sarah promised her little boy, Jimmy, that she would take him to the circus next Saturday. However, Sarah came down with the flu on Saturday. She was vomiting and she had an elevated temperature. When Sarah informed Jimmy that she was too ill to take him to the circus, he began to cry. Sarah told herself that she was "some mother, going around breaking promises to my little boy and disappointing him like that." So she got out of bed and took him anyway. When she reached the circus, Sarah fainted and had to be taken to the emergency room.

10. Tina is a graduating senior in high school. She has been a straight "A" student and is, as she puts it, "hoping and praying" to get into Harvard University." Ever since her application was filed with Harvard, she has built her day around receiving the daily mail in the hope that a Harvard acceptance letter will arrive for her. She

calls her mom several times a day from school to see if the letter has arrived. When she awakens in the morning, the first thing she thinks about is "maybe it will come today." She has told her mother that she is feeling very "nervous inside" and that she "just can't stand this pressure of waiting and not knowing."

*11. Marjorie was at a house party when she met Darrel. The two started up a conversation. Marjorie then asked Darrel if he would like to go outside with her for a walk. Darrel agreed. As they walked Darrel began to make sexual advances toward Marjorie. Marjorie asked him to stop but he persisted. He forced himself on top of her and began to pull her blouse off when another couple came by. Darrel then got off of Marjorie and fled the scene. Marjorie told herself what a "dumb fool" she was for having asked Darrel out for a walk in the first place. She said that she should never have "come on" so strong to him like some kind of "slut" and that she got just what she was asking for. Marjorie decided, therefore, not to press charges against Darrel.

12. Gene is a married, part-time graduate student working toward his master's degree in social work. He is employed full time during the day in a mental health facility and attends classes three nights each week. He is also working on a master's thesis, which he says, has gone very slow since he can't find the time to get much done on it." For the past few months Gene has had problems getting his assignments in on time. He complains that it is "hard to remember everything he has to do." He also has forgotten to attend several important staff meetings at work, for which he almost lost his job. Yesterday he was supposed to meet his wife for lunch and he never showed up or called to cancel. He told his wife, however, that "it wasn't deliberate." He just had some other things come up and he "completely forgot" about their date.

*13. Maureen is a premed student in her sophomore year at college. She presently has a 4.0 ("Straight A") cumulative grade point average. However, since Maureen began college, she has been unable to relax even on weekends, and she has trouble falling asleep at night. She tells herself that she must continue to get straight A's if she is to get into medical school. Thus, last week, when she received a "B" on her anatomy and physiology exam, she became extremely depressed. "Now," she exclaimed, "I've really blown it."

14. Jim has recently earned a Ph.D. in philosophy and plans to teach philosophy at a college or university. However, while he has had many interviews, he has been unable to get a job in his field so far. Jim is now telling himself how his Ph.D. means absolutely nothing, how he has wasted eight years of his life in school and what a total failure he has turned out to be.

15. Don and Laura were on their first date one Saturday evening. They were standing on line in a restaurant waiting to be seated when Don turned around to point out to Laura an interesting painting on the wall. However, as he turned, his elbow accidentally struck Laura in the eye. Although Laura's eye felt "poked" for a few minutes, she did not think that the incident was worth ruining their evening over. Nevertheless Don kept thinking about what a "clumsy oaf" he was, and every few minutes he kept asking Laura how her eye was feeling and he persisted throughout the rest of the evening telling her how sorry he was about it. When, at the close of their date, Don asked Laura out on another date, Laura thanked Don for the "lovely" evening but told him that she did not wish to go out with him again.

*16. Barb is a supervisor in the claims department of an insurance company. Her main objective is to become an upper level manager. In an effort to "prove herself" she has become extremely demanding of her employees. She often "flies off the handle" at them when they make even a small mistake, threatens to fire them, and watches them like a "hawk" to make sure that they make the most of their time. As a result, she has developed a poor rapport with her employees, a couple of whom have quit their jobs.

17. Alexandria is a single mother with two children, ages seven and six. She works full time as a receptionist during weekdays, and as a part-time sales clerk three evenings per week plus all day Saturday. She takes a class at a local college one evening each week; she is the secretary- treasurer of the PTO, and she is active in her church on Sundays. Her younger child has been getting into trouble in school lately and her older child has just yesterday fallen off of his bicycle and had to have several stitches in his head. In addition, the only toilet in the house has just broken, the roof is leaking, and her only car needs a new transmission. Alexandria has begun to complain about not having enough time to handle things. She says that she just can't take such "pressure" anymore.

18. Jackie is a 60-year-old widow. Her husband has recently died of cancer. She keeps telling herself that she is responsible for her husband's death because she should have made him go for a checkup sooner so that the cancer could have been discovered earlier. Sometimes she also blames her adult children, telling them that they should have had enough brains even if she herself didn't. She also tells her children that at her age there is no point in going out and trying to find anyone else and that she's probably going to end up by herself for the rest of her life.

*19. Sylvia has taken a speech class at a local college and as part of the class requirements, she was asked to prepare a speech to the class. The night before it was her turn to give the speech, Sylvia stayed up almost the entire night worrying about what a disaster it would be if she messed up. The next day, when Sylvia got up in front of the class to give her speech, her mind, as she put it, "went completely blank."

20. Fanny has had an argument with her boyfriend, Stuart. She said some things to him for which she is now sorry and hopes that Stuart will phone her so she can "get it off her chest." For two days since the argument, Fanny has refused to leave her house for

fear that she might miss his call.

21. John has recently been involved in an automobile accident. The other driver, who was severely injured, is now filing a suit against John. While John has automobile insurance, and has hired an attorney to defend him, he is afraid of what might happen if he loses the case. In fact, for the past several weeks, John has spent much of his day going over and over the case in his mind, telling himself how terrible it would be if he lost it.

22. Roy is a history professor with about four hundred students. On several occasions, Roy has forgotten to keep appointments he made with students. He has also forgotten about several faculty meetings as well as some committee meetings. Roy generally writes down his appointments on small pieces of scrap paper and leaves them in various corners of his office. Sometimes, he tapes them on a wall so that he will notice them. On a few occasions, Roy found the record of missed appointments underneath a pile of old styrofoam coffee cups. Roy believes, however, that anyone with as many students and as hectic a schedule as he would miss appointments too.

23. Jeffrie is a high school freshman. Recently, his grades have dropped from "A's to "D's" and "C's" in all of his classes. He has also started to talk to his neighbors during class instead of paying attention. In English class, the students were asked to write a short story. Jeffrie's story was about a high school student who died in a gang fight. When Jeffrie's English teacher confronted him about the decline in his grades, he responded, "It doesn't matter what grades I get; I might as well not even try. Nobody really cares how I do in school, anyway."

24. When Fred, a high school student, got back a "C" on Mr. Jenkin's biology exam, he raised his hand and accused Mr. Jenkins of giving "unfair exams" and of "stinking as a teacher." Mr. Jensen responded by calling Fred a "spoiled little baby who didn't

get what he wanted," said that he was going to give Fred "the grade he really deserved," and told him to "get the hell out" of his class and "never come back."

APPENDIX A
YOUR FAULTY THINKING PROGNOSIS: FALLACIES YOU MAY STILL NEED TO WORK ON

The following exercise will give you an idea of what fallacies you probably still need to work on after having completed a course using this text. You can validly proceed with the exercise only if you meet the following four con- ditions:

(1) You have completed the reading and done the assigned exercises for at least Chapters 1 through 6.

(2) You have used *BELIEF-SCAN* as a regular part of your course and you have, each week, for a period of at least five weeks, entered and run complete *BELIEF-SCAN* analyses of at least 5 sentences describing your own beliefs.

(3) You have kept printouts of your *BELIEF-SCAN* "Summaries of Possible Fallacies Found" (see *Operating Manual*).

(4) You have taken both pre- and post-tests of the *PRACTICAL LOGIC ASSESSMENT INVENTORY (PLAI)*. The pre-test must have been taken prior to study of the fallacies (as beginning with Chapter 2); and the post-test must have been taken *after* having read through and done the assigned exercises for at least Chapters 1 through 6

If you meet the above four conditions, then proceed as follows:

(1) On a blank sheet of paper, list *all* individual fallacies that appear on your PLAI post-test analysis. (These will be the fallacies that are listed on your analysis as fallacies that you have committed often or very often).

(2) Reading through each of your *BELIEF-SCAN* "Summaries of Possible Fallacies Found" look to see which of the fallacies you have just listed also appear on at least one of your Summaries. Place a check mark next to each such fallacies on your list.

(3) Using another blank sheet of paper, list all the fallacies on your list that have a check mark next to them. The fallacies on this

new list will be ones that, *by your own acknowledgment, you still probably have a tendency to commit often or very often.* Accordingly, label this list "**Fallacies I Know I Commit Often**"

(4) On another blank sheet of paper, list all fallacies that are listed on *two or more BELIEF-SCAN* "Summaries of Possible Fallacies Found" *but which do not appear on either of your pre- or your post-test of the PLAI. These fallacies will be fallacies that you probably still have a tendency to commit often even though you may not presently be aware of it.* Accordingly, label this list "**Fallacies I Did Not Know I May Commit Often**."

(5) Maintain the lists you have assembled in (3) and (4) above for future reference. While the accuracy and completeness of these lists will be a function of the data from which they were compiled, these lists should *together* give you an idea of what fallacies you still probably have a tendency to commit and which should therefore be given your special attention in the continued effort to improve upon your thinking habits.

APPENDIX B
SUGGESTED ANSWERS TO STARRED PRACTICE SCENARIOS

Presented below, according to Chapter, are answers to starred practice scenarios. These should be consulted only *after* you have worked through the exercises on your own. While the answers given here point to the most *specific* fallacy or fallacies committed, they are *suggested* answers only. So, if your answer does not always match the one given here, that does not *necessarily* mean that your answer is not a reasonable alternative. Note also that, while you have been asked to consider how the person who committed the fallacy in question might have avoided it--and also how the scenario might relate to your own life situation--answers provided here do not deal with these more "open-ended" issues.

Suggested Answers To Starred Practice Scenarios For Chapter Two

4. MAGNIFYING RISKS. Given what she knows about her condition, Janie is greatly exaggerating the probabilities of her needing heart surgery and of her dying prematurely. Notice Janie's use of the strong terms "probably" and "for sure."

13. UNSUPPORTED EXPLANATION. Maxine's explana- tion is not the only possible explanation. Indeed, the fact that she has not been employed for at least the past 17 years supports the explanation that the employer wanted someone with more experience. Perhaps Maxine has simply picked the possible explanation that would, if true, disturb her the most. However, this is not itself evidence that that explanation is the correct one.

18. THE SMOKE SCREEN. Polly and Patty are covering up their inability to factually support their negative attitudes toward their husbands by substituting a smoke screen of negatively charged emotional language, for example, "real turkey' and "bag of horse

manure." Notice also that some of this language is being used with sarcasm and reproach, for example, "Sweet-Old-Boy" and "little darlin."

28. HASTY GENERALIZING. Florence generalizes about all men on the basis of just two men, her husband and Vicki's. Similarly, she generalizes about all women on the basis of just two women, herself and Vicki.

30. ASSUMING THE CAUSE. Julie assumes that it was what she said that caused Mickey's illness. The mere fact that Mickey became ill after Julie remarked about how well he has been does not prove that this remarked caused his illness. Therefore, Julie irrationally blames herself for her son's illness.

44. HALF TRUTH. Sam stacks the deck against Dina by failing to acknowledge that Dina did not cause the accident. Instead, he tries to make it sound like Dina is incompetent as a driver.

Suggested Answers To Starred Practice Scenarios
For Chapter Three

3. INFERRING `IF' FROM `THEN'. Elfidge told Abigail that *if* he is to be happy living with her, *then* she needs to stop her frequent complaining. From this it does not *necessarily* follow that, because Abigail has stopped her frequent complaining (the "Then"), Elfidge will be happy living with her (the "If"). In other words, while Elfidge said that stopping the complaining was a necessary condition of his happiness, he did not say that it was a *sufficient* condition. In particular, he also wanted Abigail to continue to speak to him.

8. ARGUING IN A CIRCLE. Freda argues that Bob shouldn't have told his boss her age because it was none of his boss' business; and it was none of his boss' business, she argues, because

Bob shouldn't have told his boss her age. Freda therefore ends up justifying her objection by repeating it.

13. INCONSISTENCY. On the one hand, John says that homosexuals are sinners because they do what the Bible forbids. On the other, he thinks that his having committed adultery does not make him a sinner even though the Bible forbids adultery. However, if doing whatever the Bible forbids makes one a sinner, then John would also be a sinner. He can't have it both ways.

18. FALLACY OF THE PART. Dave assumes that because Morris Memorial, considered as a whole, is not a good hospital, then one of its staff members, Dr. Loomis, is also not good. However, what is true of a whole need not be true of a part, and Dr. Loomis could still be a very good (competent) physician.

23. GRAY AREA FALLACY. While it may be difficult to say just where social drinking leaves off and alcoholism begins, the fact that there is a gray area between the two does not mean that there are not any clear cases of either. There are some people who are clearly alcoholic, and Joe may well be one of them.

Suggested Answers To Starred Practice Scenarios
For Chapter Four

3. SELF-EVIDENCE FALLACY. James takes it as self-evident that a man must not cry. However, this is no part of the meaning of "man." A man does not shed his gender through the act of crying! James is therefore irrationally disturbing himself over the imagined loss of his manliness.

8. OPPOSITE MEANING. Carla says she doesn't mind her boss' smoking when she really does. Instead of being direct about her health problem concerning inhaling smoke, she relies on cues such as coughs and verbal slips to convey her true thoughts. Her boss,

however, does not pick up on these cues, and Carla ends up consenting to a situation that could prove harmful to her health.

11. VAGUE TERMS. While Shelly finds Paul to be "very nice," she dismisses going out with him because he has ideas that are "strange," "weird," "peculiar," "way out." Moreover, she favors someone who is "more down to earth," someone "more like the way [she] is." Yet the actual meaning of these terms is unclear. By clarifying them, she may be be in a better position to make a decision about Paul. For instance, what is "very nice" about him? What is "weird" about his ideas? What does it mean to be "down to earth"? Shelly might also ask Stacey to consider what she means by a "real catch." Presumably Bob is not (literally) a big fish!

16. PERSONIFYING. Bert is giving human traits to inanimate objects such as his cars, house, job, and "things" in general. For example, "things" cannot "get you by the throat." Nor can one be made a "slave" by one's house. Nor do cars conspire against their owners. Nor are jobs "real bitches." By speaking in this way, Bert prevents himself from realizing the control he can, indeed, exercise over the things in his life.

Suggested Answers To Starred Practice Scenarios For Chapter Five

5. PARROTING. Sandy is simply repeating back what her parents have said about her without examining the realities of her situation. Because of this, Sandy fails to take into account her professor's relevant observations about her classroom participation and essays.

7. PERSONAL ATTACK. Mort fails to consider Howard's advice because Howard has been divorced. By doing this, Mort appears to be judging Howard personally rather than considering the appropriateness of the advice itself.

14. MISUSE OF PITY. Terry's father has reversed his decision about her viewing "R-rated" movies out of pity for her and without due consideration of relevant facts, such as why the movie has been rated "R" in the first place. Terry's father could have been sensitive to Terry's feelings and still based his decision on relevant considerations.

33. ARGUMENT FROM IGNORANCE. From the fact that Maureen is unable to prove that Jim is after her money, it does not follow that he is, indeed, after her money. However, Milley has convinced Maureen that this is the case.

36. APPEAL TO TRADITION. Dr. Porterhouse is insisting on tradition for the sake of tradition. He fails to realize that in this instance, the practice of blindly following tradition does not take into account his son's capabilities or aspirations. Dr. Porterhouse's insistence on tradition might, indeed, jeopardize not only his son's future happiness, but could also contribute to a strained father-son relationship.

39. MISUSE OF AUTHORITY. Pam has mistakenly assumed that, because her friend Sharon is a physician's receptionist, that she too is an authority on physical ailments. Although Sharon may, in fact, perform well in her capacity of receptionist, this does not, in any way, give her the expertise to dispense medical advice.

Suggested Answers To Starred Practice Scenarios
For Chapter Six

7. AWFULIZING; SELF-DAMNATION. Lucille is exaggerating the negative significance of her husband's leaving her for another man. It appears that her husband is homosexual and is not comfortable in a heterosexual relationship. Since Lucille is not responsible for her

husband's homosexuality, she cannot sensibly
8. blame herself for it. Nor is it as bad as she thinks since she can still seek out new relationships in the future. Moreover her husband may achieve greater happiness through a homosexual relationship.

10. THE "THOU SHALT UPSET YOURSELF" FALLACY. Roselyn takes on a duty to upset herself over her son's conduct and condemns her husband for a breach of this duty. In her view, she must remain upset so long as the problem in question persists. However, the imposition of such a "duty" is groundless. Moreover, if Roselyn continues to live by such an ethic, she is likely to endure a very unhappy existence.

13. THE "I JUST CAN'T HELP THIS FEELING" FALLACY. Craig fails to take any responsibility for his having gotten upset. While Tanya's remarks were intended to be offensive, Craig's response to those remarks was largely a function of the beliefs he entertained about them. For example, if he told himself how awful it was to be called "boring" and how a wife should never say such things, then he might well be upset. But Craig did not have to tell himself such things in the first place. Notice his use of the responsibility hiding terms ""You really hurt me," "Because of you," "you're going to give me a heart attack."

17. DEMANDING PERFECTION (IN THIS IMPERFECT WORLD); SELF- DAMNATION. Dwayne self-defeatingly commits the fallacy of Demanding Perfection by demanding that he never again commit the fallacy of Demanding Perfection. He then goes on to negatively rate himself ("Now I feel like such a fool") thereby also committing the fallacy of Self-Damnation.

27. POISONING THE WELL. Alice allows herself to be manipulated and "poisoned" away from her studies by Pauline's use of strong negative emotional language, for instance, "such a bore," "dumb test," and "bookworm."

Suggested Answers To Starred Practice Scenarios
For Chapter Seven

2. LEARN HOW TO SAY NO. Since Pat has earned a solid reputation among his colleagues for doing whatever they ask of him, they are likely to keep on asking him. However, if Pat is to advance his own interests instead of merely the interests of others, he will have to learn how to say no to others' requests. Moreover, he need not apologize for refusing what he is not obligated to do. For example, if he is already on more than the required number of committees, he is justified in simply pointing this fact out when asked to serve on further committees on which he would prefer not serve.

7. BE ESPECIALLY ON THE LOOKOUT FOR DAMNING THOUGHTS ABOUT YOURSELF. Sonia blames herself for her husband's having left her. Moreover, when another man tries to get close to her, she turns him away, perhaps because she has already told herself that she is unworthy of that man also. Moreover, despite the fact that her husband was the one who had the affair, she makes him out to be faultless (and so commits the "Terrific" Fallacy). According to Sonia, she was the one who made mistakes, not him; and she was the one who didn't deserve someone as good as him. Sonia fails to realize that her husband was not faultless and that she is unwarranted in Damning herself for her divorce.

10. PLEASE YOURSELF; SELF-EVIDENCE FALLACY. Kathy puts off pleasing herself because she thinks that to do otherwise would make her a selfish person. However, there is no "self-evident" law of nature requiring that people sacrifice their own happiness for the interests of others. Kathy has committed the Self-Evidence Fallacy by supposing that there is such a requirement, and that, therefore, she must be a martyr!

13. AVOID DILEMMA THINKING. Elias poses a dilemma for the couple: either they have a child first or they buy a house, but

not both. But regardless of which of these two options they take, they will serious impair their opportunity for doing the other. This dilemma, however, does not appear to be a real one. First, there seem to be other options. For example, they could get a loan or purchase a less expensive house and try to have a child in the meanwhile. (Does Elias even know how long it will take to conceive, as though he and Geraldine could have a child in the same way that they could set a closing date?) But even if they are not willing to try for both, Elias appears to be jumping onto a Slippery Slope by exaggerating the financial results of buying a house. Would they really then be incapable of supporting a child or would it just be more difficult? Similarly, would having a child exclude them from buying any house for years? Finally, Elias might have better avoided dilemma thinking altogether, since such thinking is likely to lead to inactivity and Decision By Indecision. A more constructive approach might have been to set priorities and to act upon them, without trying to turn their case into a dilemma. For example, they might simply have decided that a child was more of an immediate priority and that they would reassess their financial situation after they had the child.

Suggested Answers To Starred Practice Scenarios
For Chapter Eight

5. DO WHAT YOU CAN ABOUT YOUR PROBLEM; DEPRESSION INVOLVING SLIPPERY SLOPE/ AWFULIZING. Instead of constructively involving himself in potentially helpful physical therapy, Jerry insists on dwelling on how horrible his situation is. By exaggerating the negative consequences of even trying to help himself ("I'll never be the same no matter what I do"), and then, by Awfulizing about those consequences, Jerry makes himself depressed.

11. GUILT INVOLVING DEMANDING PERFECTION AND SELF- DAMNATION. Margorie's unrealistic demand on herself causes her to blame herself for not being able to anticipate and

control the actions of another person. In addition, her Self-Damnation leads her to experience guilt and self-downing behavior. As a result, Marjorie irrationally absolves Darrel of guilt for the attack and decides not to press charges.

13. ACCEPT YOURSELF AS AN IMPERFECT PERSON; ANXIETY AND DEPRESSION RESULTING FROM DEMANDING PERFECTION, AWFULIZING AND SLIPPERY SLOPING. Maureen's insistence on perfection in her studies has been a source of continued anxiety in her life. Moreover, Maureen's committing the Slippery Slope/Awfulizing fallacies has caused her to irrationally exaggerate the negative consequences of a grade of "B" on an exam, leading her to the "horrible" conclusion that she will not be accepted into medical school. Thus, Maureen's current depression results from a combination of unrealistic beliefs and demands. Had she realistically appraised the situation, Maureen would be able to recognize that it is irrational to seek perfection of herself, that her exam grade is unlikely to cause her rejection from medical school and that even if she were to be rejected from medical school, it would not be the end of the world.

16. ANGER INVOLVING DEMANDING PERFECTION AND APPEAL TO FORCE. Barb is Demanding Perfection not only of herself but of her employees as well. By embracing the belief that her employees' performances are a reflection of her own self-worth, Barb has unrealistically attempted to monitor every detail of their work. Barb's anger at her employees results from her irrational belief that any mistakes they make is going to have a negative impact on her prospects for promotion. Barb's threats to fire those who work for her have, in fact, compounded her stress as some have already quit their jobs.

19. ANXIETY DUE TO AWFULIZING. Sylvia's pre- occupation with giving her speech is based on an irrational belief about the devastating consequences of a poor performance. If Sylvia had realistically assessed her situa- tion, she would have concluded that

even a poor per- formance would not mean disaster and that this event did not have to result in a bleak future for her. It appears, as well, that Sylvia's irrational beliefs have led her into such a state of anxiety that the very situation that she feared has resulted.

FURTHER READING

Allport, Gordon W. *The Nature of Prejudice*. Reading, Mass.: Addison Wesley Publishing Co., 1979.

Arthur, John, Ed., *Morality and Moral Controversies*. 3rd Ed. Prentice-Hall, 1993.

Baum, Robert, Ed., *Ethical Arguments For Analysis*. New York: Holt, Rinehart, and Winston, 1979.

Beck, A. T., A. J. Rush, B. F. Shaw, and G. Emery. *Cognitive Therapy of Depression*. New York: Guilford Press, 1979.

Berne, Eric. *Games People Play*. New York: Ballantine Books, 1964.

Burns, David D. *Feeling Good*. New York: Penguin Books, 1981.

Chaffee, John. *Thinking Critically*. Boston: Houghton Mifflin, 1985.

Cohen, Elliot D., Ed., *Philosophers at Work: An Introduction to the Issues and Practical Uses of Philosophy*. New York: Wadsworth, 2000.

Cohen, Elliot D. *What Would Aristotle Do? Self-Control Through the Power of Reason*. Prometheus Books, Amherst, NY, (2003).

Copi, Irving C. *Informal Logic*. New York: Macmillan Publishing Co., 1986
.
Diestler, Sherry. *Becoming A Critical Thinker: A User-Friendly*

Manual. New York: Macmillan, Inc., 1994.

Ellis, Albert and Robert A. Harper. *A New Guide to Rational Living*. No. Hollywood, Calif.: Wilshire Book Co., 1975.

Ellis, Albert, *Overcoming Destructive Beliefs, Feelings, and Behavior*. Prometheus Books, Amherst, NY, 2001.

Engel, S. Morris. *With Good Reason: An Introduction to Informal Logic*. 4th Ed.. New York: St. Martin's Press, 1990.

Gold, Steven Jay, ed. *Moral Controversies: Race, Class, and Gender in Applied Ethics*. Belmont, Calif.: Wadsworth Publishing Co., 1993.

Kahane, Howard. *Logic and Contemporary Rhetoric: The Use of Reason in Everyday Life*. 5th Ed. Belmont, Calif.: Wadsworth Publishing Co., 1988.

Martin, Mike W. *Everyday Morality: An Introduction to Applied Ethics*. Belmont, Calif.: Wadsworth Publishing Co., 1989.

May, Larry and Shari Collins Sharratt, eds. *Applied Ethics: A Multicultural Approach*. Englewood Cliffs, N.J.: Prentice-Hall, 1994.

Mayfield, Marlys. *Thinking for Yourself: Developing Critical Thinking Skills Through Writing*. 2nd Ed. Belmont Calif.: Wadsworth Publishing Co, 1991.

Moore, Kathleen Dean. *Reasoning and Writing*. New York: Macmillan Publishing Co., 1993.

Moore, Noel Brooke and Richard Parker. *Critical Thinking*. 3rd. Ed. Mountain View Calif.: Mayfield Publishing Co., 1992.

Moore, W. E., Hugh McCann, and Janet McCann. *Creative and Critical Thinking*. 2nd Ed. Boston: Houghton Mifflin, 1974.

Ruchlis, Hy. *Clear Thinking: A Practical Introduction.* Buffalo: Prometheus Books, 1990.

Rudinow, Joel and Vincent E. Barry. *Invitation to Critical Thinking*. 3rd. Ed. Fort Worth: Harcourt Brace, 1994.

Ruggiero, Vincent. *The Art of Thinking: A Guide To Critical and Creative Thought*. 2nd Ed. New York: Harper & Row, 1988.

Seech, Zachary. *Open Minds and Everyday Reasoning*. Belmont, Calif.: Wadsworth Publishing Co., 1993.

Soccio, Douglas J. and Vincent E. Barry. *Practical Logic: An Antidote for Uncritical Thinking*. 4th Ed. Fort Worth: Harcourt Brace, 1992.

Solomon, Robert C. *The Passions: The Myth and Nature of Human Emotion*. New York: Anchor Books, 1977.

Walen, Susan R, Raymond DiGiuseppe, and Windy Dryden. *A Practitioner's Guide to Rational-Emotive Therapy*. 2nd Ed. New York: Oxford University Press, 1992.

Zechmeister, Eugene B. and James E. Johnson. *Critical Thinking: A Functional Approach*. Belmont, Calif.: Brooks/Cole Publishing Co., 1992.